RECOGNIZING EUROPEAN MODERNITIES

For over a century Europe has been characterized by a multiplicity of capitalist modernities. At any moment, each country possesses its own distinctly modern qualities which are partly shaped through interrelationships with other countries. Each European commodity society has experienced successive, but differently overlapping, periods of industrial modernity (large-scale factories and urban growth), high modernity (social modernization promoted by social engineering) and hypermodernity (the acceleration of modernity, yielding new circumstances and sensibilities). Investigating any part of contemporary hypermodern Europe thus requires that it be brought into constellation with its industrial and high modern past.

Recognizing European Modernities explores a century of civilization through a critical examination of the extreme case of Sweden. Using montage – relayering multiple pasts and on-going presents – the book challenges the contemporary obsession with "postmodernity," demanding a deeper, more informed understanding of the extended danger characteristic of the European present.

The author visits three spectacular spaces: the Stockholm Exhibition of 1897, the Stockholm Exhibition of 1930, and the Globe, a contemporary multi-purpose arena. Analysis of these pivotal spaces reveals the on-going process of modernization as new forms of consumption are repeatedly entangled in changing discourses of power, both of which become reworked and translated into cultural politics.

RECOGNIZING EUROPEAN MODERNITIES

A Montage of the Present

•

ALLAN PRED

London and New York

First published 1995
by Routledge
11 New Fetter Lane, London EC4P 4EE

Simultaneously published in the USA and Canada
by Routledge
29 West 35th Street, New York, NY 10001

© 1995 Allan Pred

Typeset in Garamond and Franklin Gothic by
Solidus (Bristol) Limited,
Printed and bound in Great Britain by
Butler & Tanner Ltd, Frome and London

All rights reserved. No part of this book may be reprinted or
reproduced or utilized in any form or by any electronic,
mechanical, or other means, now known or hereafter
invented, including photocopying and recording, or in any
information storage or retrieval system, without permission in
writing from the publishers.

British Library Cataloguing in Publication Data

A catalogue record for this book is available from the British Library

Library of Congress Cataloging in Publication Data

A catalogue record for this book has been requested

ISBN 0–415–11904–9
0–415–12136–1 (pbk)

It isn't that the past casts its light on what is present or that what is present casts its light on what is past; rather, an image is that in which the Then [and There] and the [Here and] Now come together into a constellation like a flash of lightning.

WALTER BENJAMIN (1972–1989), VOL. V

CONTENTS

Acknowledgments
9

For(e) Montage:
Swedish Modern and the Recognition of
European Modernities
11

CHAPTER ONE
Spectacular Articulations of Modernity:
The Stockholm Exhibition of 1897
31

CHAPTER TWO
Pure and Simple Lines, Future Lines
of Vision:
The Stockholm Exhibition of 1930
97

CHAPTER THREE
Where in the World Am I, Are We?
Who in the World Am I, Are We?:
The Glob(e)alization of Stockholm,
Sweden
175

After Montage: Or What?
257

Bibliography
265

Index
283

ACKNOWLEDGMENTS

•

What we [as cosmopolitan intellectuals] share as a condition of existence...is a specificity of historical experience and place, however complex and contestable they may be, and a world-wide macro-interdependency encompassing any local particularity. Whether we like it or not, we are all in this situation.

PAUL RABINOW
(1986), 258

The form and content of the critical human geography which suffuses this book owes as much to the all too long dead Walter Benjamin, as to anybody. But none of it would have been possible without the many years of interaction, influence and inspiration provided by my departmental colleagues Michael Watts and Dick Walker (and, more recently, Michael Johns); without the transdisciplinary seminars on the intermeshed spatiality of culture, power and knowledge jointly taught with Paul Rabinow; and without, not least of all, the difficult questions and intellectual excitement supplied by a long line of truly remarkable graduate students, including, most recently, Katharyne Mitchell, Victoria Randlett, Elizabeth Vasile, Jim Proctor, Rick Schroeder, Eric Hirsch, George Henderson, Susan Pomeroy, Mark O'Malley, Jennifer Jones, James McCarthy, Sharad Chari, Kathryn Caldera, Lisa Hoffman, Marta Gutman, Tad Mutersbaugh, Jorge Lizàragga, Kathy McAfee and Grey Brechin. Nor would the innovative and critical qualities of this book have been possible without my periodic, but all too infrequent encounters, with elsewhere based human geographers, most especially Derek Gregory, Nigel Thrift, Neil Smith, Cindi Katz and Susan Christopherson.

The contents of this book cannot be separated from the fact that I have spent nearly 40 per cent of my life in Sweden since 1960, a circumstance allowing me to function as an inside outsider – as an observer practically and socially enmeshed in the richness of everyday life rather than simply as a somewhat removed visiting researcher. As Chapter 3 makes evident, this has enabled me not only to make use of an intense daily exposure to the various mass media; but also to stockpile – through unstructured informal conversations participated in and overheard in public places – the viewpoints of countless people of both genders who vary widely in terms of age, occupation, class background and political allegiance. These numerous women and men, known and unknown, but never here identified, are to be thanked for enriching my knowledge and understanding.

As a consequence of my numerous extended Swedish sojourns I have enormous, long-standing intellectual and personal debts to Gunnar Olsson, Torsten Hägerstrand and Sven Nordlund. I have also time and again benefited from contacts with Orvar Löfgren and Ulf Hannerz, as well as with Roger Andersson, Sune Berger, Bo Malmberg and the entire group of young scholars at Uppsala University's Department of Human Geography. Generous support from, and residence at, the Swedish Collegium for Advanced Studies in the Social Sciences throughout 1991 enabled me to set this project in motion. Oscar Reuterswärd, the noted artist and art historian, provided me with a pivotal and memorable afternoon of recollections, interpretations and encouragement. Without the exceptional efforts of Barbro Klein and her students at the University of Stockholm's Department of Ethnography, much of the ethnographic material appearing in Chapter 3 would not have been available. Lengthy and insightful commentaries on that chapter were provided by Lena Sommestad, as well as Orvar Löfgren and Gunnar Olsson. Eva Rudberg provided helpful tips at the outset of my explorations for Chapter 2. Lasse Johansson and Åke Abrahamsson were extremely thoughtul and suggestive in helping me to exploit the archival holdings of the City Museum of Stockholm. Bo Theen of the Swedish Museum of Architecture went out of his way to assist me in assembling photographs. The editors of *Geografiska Annaler* and *Nordisk Samhällsgeografisk Tidskrift* kindly granted me permission to further develop published articles that now appear as Chapters 1 and 2. Anne Unge, of AB Stockholm Globe Arena, and the cultural page staff of *Dagens Nyheter* made exceptions so as to grant me unrestricted access to the press clipping collections of their respective organizations. Above all, I am indebted to Hjördis, who not only has nourished me with love, friendship and care throughout this project, but also has provided invaluable daily lessons in the nature of Swedish modern. Erik and Michele, as always, have helped me keep my feet on the ground.

As specified at the end of Chapters 1–3, numerous photographs and art reproductions appear by courtesy of *Stockholms Stadsmuseum* (the Stockholm City Museum), *Arkitekturmuseet* (the Swedish Museum of Architecture) and *Norrköpings Konstmuseum* (The Norrköping Art Museum).

FOR(E) MONTAGE: SWEDISH MODERN AND THE RECOGNITION OF EUROPEAN MODERNITIES

●

Method of this work: literary montage. I have nothing to say, only to show.	WALTER BENJAMIN (1982), 574, as quoted in Susan Buck-Morss (1989), 73

When Benjamin praised montage as progressive because it "interrupts the context into which it is inserted," he was referring to its destructive critical dimension.... But the task of the Arcades project was to implement as well the constructive dimension of the montage, as the only form in which modern philosophy could be erected.

    [I]f Benjamin threw the traditional language of metaphysics into the junkroom, it was to rescue the metaphysical experience of the objective world, not to see philosophy dissolve into the play of language itself.

SUSAN BUCK-MORSS (1989), 77, 223

[I]n both cases [Foucault's archeology and Benjamin's Arcades Project (Das Passagen-Werk)][1] "archeology" is not so much an excavation, bringing buried or hidden objects to the surface, as a way of showing the particular – anonymous dispersed – practices and the particular – differentiated, hierarchized – spaces through which particular societies make particular things visible.

DEREK GREGORY (1991), 36

[C]onsumption [in twentieth-century Sweden may be seen] as everyday praxis, as Utopia and as an ideological battlefield.

ORVAR LÖFGREN (1990b), editorial preface, 5

ALAN TOMLINSON (1990a), 30	[T]he politics of consumption are central to the dynamics of social and cultural change in contemporary capitalism.
DANIEL MILLER (1987), 215, 177	Mass goods represent culture, not because they are merely there as the environment within which we operate, but because they are an integral process of objectification by which we create ourselves as an industrial society: our identities, our social affiliations, our lived everyday practices. Material forms ... lend themselves admirably to the workings of both ideological control and uncontested dissent.
PAUL WILLIS (1990), 21	People bring living identities to commerce and the consumption of cultural commodities as well as being formed there. They bring experiences, feelings, social position and social memberships to their encounter with commerce. Hence they bring a necessary symbolic pressure, not only to make sense of cultural commodities, but partly through them also to make sense of contradiction and structure as they experience them [in everyday life], and as members of certain genders, races, classes and ages.
ORVAR LÖFGREN (1992)	[T]he reworking of the mass produced occurs within distinct cultural frameworks which vary not only between gender, class and generation, but also between nations.

Swedish (high) modern. – An image of Swedish goods broadcast abroad as well as at home. Especially since the Stockholm Exhibition of 1930, a term synonymous with the clean and often graceful lines – "the tasteful reservation" – of Swedish functionalism, of Swedish high-modern design.[2] Swedish furniture. Swedish glassware. Swedish china and kitchenware. Practical Swedish handicrafts.... Swedish (high) modern(ity) – a matter of consumption.

Swedish (high) modern. – A question of norms as well as forms.[3] A design aesthetic that was inseparably interwoven with a national project of social modernization. With the establishment of a new social order. With social engineering and the construction of both a new "man" and a new national identity. An aesthetic that paradoxically linked new forms of collectivism with individualism. An aesthetic that was one with discourses that spoke of solidarity, of social responsibility, of individual freedom, of pride in Swedish industrial accomplishments. Acceptance of the present, with its simple-lined technological marvels, and optimism about the future. An agenda serving corporate capital as well as the Social Democrats.... Swedish (high) modern(ity) – a matter of (would-be) hegemonic discourses.

Industrial modernization arrived in Sweden a half-century before the project of social modernization was given concrete expression at the Stockholm Exhibition of 1930.

Swedish (industrial) modern. – The initial appearance and expansion – most notably in Stockholm – of forms of competitive industrial capitalism already characteristic of many urban centers elsewhere in Western Europe and North America. An erratically paced surge of economic restructuring concentrated in the 1880s and 1890s (although it did reach some Stockholm production sectors as early as the 1870s).[4] A process of modernization during which manufacturing output became increasingly concentrated in large-scale factories, thereby undercutting the viability and relative importance of artisan and small workshop production. A superimposition of new practices and power relations by local capitalists responding to, and contributing to, transformations in the European world economy and the international division of labor. A period during which manufacturers generated – and depended upon – the further spread of commodity fetishism, the widespread diffusion of fads and short-lived fashions, the marketing of mass-produced goods to a mass market. A set of production *and consumption* developments which was dazzlingly manifested at the Stockholm Exhibition of 1897.... Swedish (industrial) modern(ity) – as much as anything, a matter of consumption.

Swedish (industrial) modern. – A question of new technologies, stricter labor discipline and widespread periodic poverty. A question of intensified work-site and place-specific class tensions as well as new forms of production and consumption. The emergence of discourses which denied or attempted to cover over these and other social tensions. Discourses which legitimized through messages equating social progress with the wonders of technology, through representations of new commodities to come and an ever better future, through rhetorics of national pride and social harmony.... Swedish (industrial) modern(ity) – a matter of (would-be) hegemonic discourses.

[The] common theme [of Kurt Schwitters' "Merz" paintings] was the [modern] city as compressor, intensifier of experience. So many people, and so many messages: so many traces of intimate journeys, news, meetings, possession, rejection, with the city renewing its fabric of transaction every moment of the day and night, as a snake

ROBERT HUGHES
(1991), 64

casts its skin, leaving the pattern of the lost epidermis behind as "mere" rubbish.

    The practices and experiences of everyday life during the modernity of late nineteenth-century European and North American industrial capitalisms were characterized in geographically specific ways by encounters
with a "ruthless centrifuge of change,"[5]
with the transitory and the apparently fragmented,
with an incessant spectacle of the new,
with repeated material evidence of creative destruction,
with relentlessly expanding commodification,
with fleeting forms of consumption
      and a dreamworld of commodity fetishism,
with constant changes in employment opportunities
      and circumstances.
All of which were repeatedly subject to popular, artistic and intellectual reworkings, to symbolic discontent and cultural struggle. The practices and experiences in question were further characterized by encounters
with ephemeral and disjointed social contacts,
with a jumble of local and far-flung distant images produced by ever
      more rapid and far-reaching information networks,
with the persistent spread of bureaucratic rationality
      and its iron-cage rules,
with powerful new technologies and new technologies of power,
with reconfigured articulations between the local, the national
      and the global,
with reconfigured mediations between the discourses of gender,
      class and "race" or ethnicity.
All of which were repeatedly subject to popular, artistic and intellectual reworkings, to symbolic discontent and cultural struggle.

In short,
these were encounters
with plural complexities,
with shifting heterogeneities, and
with shockingly new meanings
    that repeatedly dislocated and displaced,
    that confirmed rupture,
    that invited or demanded cultural contestation

>   by calling the meanings
>
> >   of locally preexisting practices and social relations
> >   into question,
>
> and thereby called individual and collective identities
>
> >   into question.

> This being the case,
> then the everyday life and experiences of the here and now,
> >   of the 1990s in specific places,
> however clearly distinctive and dramatic they may seem,
> however radically altered they may have become
> >   by post-Fordist,
> >   >   post-colonial or
> >   >   post-cold war circumstances,
> are best characterized as modernity magnified,
> as capitalist modernity accentuated and sped up,
> as *hyper*modern,
> not postmodern.

———

Swedish (hyper)modern. – By the end of 1992, if not earlier, Sweden was a country adrift, its population unanchored from major elements of the long taken-for-granted. Its capital, Stockholm, a metropolis economically and socially unmoored, caught in the swirling tides and eddies of (further) economic globalization. Sweden, the model welfare state, was no longer the model welfare state. Sweden, the most high-modern of all high-modern states, was no longer committed to high modernity. Sweden, the founder of the "middle way," the pathbreaker of capitalism with human dignity, the pioneer of capitalism with large (state-imposed) doses of social justice and egalitarianism, the trailblazer of intimate relations between big capital and big government, was no longer following the "middle way." Sweden, the quintessence of social democracy, the land of the "People's Home" – where all, the weak as well as the strong, were to feel welcome, secure and sheltered – was no longer a home to social democracy. Stockholm, still a city of arrestingly beautiful vistas and physical seductiveness, was no longer beautiful or seductive in other terms, was no longer the capital of the model welfare state, of the "middle way," of social democracy. Now Stockholm was instead the capital of a country whose political leaders were committed to a doctrine of "the free market," lower taxes and human deliverance or "self-realization" via greater freedom to consume. Not coincidentally, for three years the city had possessed a new dominant feature on its landscape which was not simply an imposing edifice, but also a massive monument to consumption – "The Globe," at one and the same time an ever-presence like global forms of capital, a structure in which to consume global entertainment commodities, and a device for marketing Stockholm as an international center of investment opportunities, corporate capital, banking and tourism.

Swedish (hyper)modern. – Disarray. Instability. Insecurity. Turmoil. Disillusionment. Crisis. These were the conditions that best characterized the economy, politics and everyday life of the country as a whole; conditions that were in intense evidence in the country's principal urban center. These were the conditions that summarized the experiences of a population caught up in the throes of *hyper*modernity, of capitalist modernity accentuated and sped up. The transitory, ephemeral, fragmented dimensions of everyday life were ever more so. The economy was in a shambles, brought to a sorry state by, among other things, a period of unbridled domestic speculation and the global interdependence of recession-wracked national and regional capitalisms. The State was confronted by a rapidly burgeoning deficit and fiscal crisis. The country's gross national product was falling without any sign of let up. One after another, like a long line of dominoes, factories and other business establishments were either being totally closed down or having their scale of operations severely cut back. Unemployment was achieving levels unknown since the depression. Major banks, reeling from enormous credit losses, were forced to turn to the State for assistance. The Social Democrats were not only no longer in power (having been displaced in the fall of 1991 by a conservative-led coalition of self-styled "bourgeois" parties), but, with a rightward shift in their ideology, no longer recognizable as social democrats. In many municipalities and provinces authorities were quickly moving to pare back or radically restructure welfare and cultural institutions, either drastically reducing their allocations, fully privatizing them, or contracting out their operation to private firms. The once famous open door was being slowly shut – a new hard-line, hard-nosed, increasingly restrictive position was being taken toward the admission of refugees. The very nature of the Swedish State was up for grabs as the government moved through the formal procedures of applying for membership in the European Community and an eventual European Union, as leading business figures, politicians and economists sought to convince the public of the wisdom of such a dramatic move. Swedes in general, Stockholmers in particular, no longer recognized either the world in which they lived, or themselves. Many, among other things, had lost their confidence in the honesty of those positioned above, had lost their belief in solidarity and equality, had lost their self-satisfied conviction that Sweden was the best place in the world to live, a relative heaven on earth, an exceptional country "immune to the crises affecting the world at large."[6] The dis-ease of uncertainty was pandemic. Identity crises abounded.[7]

Swedish (hyper)modern. – What was occurring, what was being experienced, was no ordinary disjuncture, no run-of-the-mill perturbation, not just another downturn in the "business cycle" inevitably wrought by the logic of capital. It was not just another modern jolt of the new, but a severe rupture. It was a number of shocks in constellation. A hypershock. The fast fade of an era. It was the long dominant becoming no longer dominant, being replaced, becoming residual, placed in a position from which it could not re-emerge – even with a Social Democratic re-election – unless in a new, highly transformed guise. It was a set of key turning points in the place-specific, on-the-ground histories of Swedish modernity – in the multitude of (geographical hi)stories of which Swedish modern is comprised. It was the unravelling of a commitment to (social democratic, state-bureaucratic) social engineering, to a politics and an aesthetic of pure and simple lines that dated back to the 1930s. It involved the finalizing of a drastic redefinition of the often uneasy and overtly antagonistic but practically co-operative and power-reinforcing relations between internationally successful Swedish capital, the Social Democratic Party and the Swedish Confederation of Trade Unions (LO) as well as its white-collar counterparts. To many it seemed that where the familiar and taken for granted once were, there was little more than shadows in a near void. In that near void a reconstructed set of (would-be) hegemonic discourses hollowly echoed – praising the operation of "market forces," proposing such forces as the solution to all social problems, promising a new consumption paradise.

The spectacle, and thereby the spectacular space, as (supposed) total hegemony:

> *The spectacle is the moment when the commodity has attained the "total occupation" of social life. Not only is the relation to the commodity visible but it is all one sees: the world one sees is its world.*

GUY DEBORD (1983)

> *The hegemonic struggle for power takes place on and across an already constituted field, within which the identities and positions of the contesting groups are already being defined but are never fixed once and for all. Hegemonic politics always involves the ongoing rearticulation of the relations between, and the identity and positions of, the ruling bloc[s][8] and the subordinate fractions within the larger social formation.*
>
> *Hegemony is organized around an explicitly defined national*

LAWRENCE GROSSBERG (1992), 245, 247

project of restructuring the social formation, a project which mobilizes the struggles of popular culture and daily life. The project is, finally, an attempt to reconstruct daily life and its relationship to the social formation. But it always remains a project, rarely completed, always changing in relation to its changing circumstances and always needing further work.

RAYMOND WILLIAMS (1977), 110, 112

[H]egemony is a lived system of meanings and values – constitutive and constituting – which as they are experienced as practices appear as reciprocally confirming.... It is, that is to say, in the strongest sense a "culture," but a culture which has also to be seen as the lived dominance and subordination of particular classes [or groups].

A lived hegemony is always a process.... It is a realized complex of experiences, relationships, and activities, with specific and changing pressures and limits. In practice, that is, hegemony can never be singular.

MICHAEL RYAN (1989), 18–19

One could argue ... that hegemony is in fact itself a kind of resistance, a means of securing the identities of property and of political power against very real internal threats that are always potentially active in a system of economic subordination founded on a radical difference or inequality.... The biggest threat, of course, is quite simply that anyone who is subordinated might reverse the line of force that maintains subordination.... [H]egemony has to be understood as the containment of a force that bears the potential of tipping the balance of power.

MICHAEL J. WATTS (1991), 8, paraphrasing M. Denning (1990)

Rather than capitulating to a view of mass culture as something which is simply a form of class control or alternatively an authentic form of subversion ... [Stuart Hall, Fredric Jameson and Paul Willis] relocated these polarities [containment and resistance, reification and utopia] by suggesting that cultural creation under capitalism was divided against itself.

PERRY ANDERSON (1990)

[Raymond Williams gave Gramsci's notion of hegemony] a characteristic twist by emphasizing the continual process of adjustment needed to secure any political or cultural hegemony above, and its perpetual failure, as an inherently selective definition of reality – to exhaust the meanings of popular experience below.

> [I]t is because no meaning is actually fixed that there is a space in which hegemonic struggle can take place.
>
> ERNESTO LACLAU
(1988), 249

> Society is a battlefield of representations, on which the limits and coherence of any given set are constantly fought for and regularly spoilt.
>
> T.J. CLARK
(1984), 6

> Hegemony is hard work.
>
> STUART HALL
adage as cited in
Michael J. Watts
(1991), 11

One way in which to approach a century of Swedish modernity, to interrogate both the Swedish (hyper)modern and those Swedish moderns which speak to it, to write a Stockholm-centered set of (geographical hi)stories of the Swedish present, is to focus on three spectacular spaces of consumption: the Stockholm Exhibition of 1897, where a new (dream)world of everyday consumption items was presented as spectacle; the Stockholm Exhibition of 1930, where a new aesthetic and politics of consumption was promoted via spectacle; and "The Globe," where the spectacle is made available as an everyday item of consumption.

Each of these spectacular spaces was physically and symbolically constructed at a pivotal (dis)juncture in Swedish modernity. None of these spectacular spaces was — or is — a mere reflection of its place and time, a simple mirror of then and there — or here and now — conditions. Each of these spaces was a crucible in which the new crystallized out of the ongoing. Each of these spectacular spaces emerged out of a series of ongoing structuring processes and — by virtue of being spectacular and hence attention-grabbing — dramatically contributed to the perpetuation of these processes, to the transformation and reconstitution of situated practices, power relations and forms of individual and collective consciousness. Each of these spectacular spaces was — or remains — dedicated to the consumption of (would-be) hegemonic discourses as well as to the consumption of commodities. Each of these spectacular spaces was — or remains — a site of cultural struggle as well as a site of commodity promotion. The construction and operation of each of these spectacular spaces was — or is — entangled in (would-be) hegemonic discourses, in discourses which were — or are — meant
to maintain, restructure or consolidate domination
    by means of popular consent rather than coercion,
to legitimate or extend existing power relations,
to ensure or produce a new relative stability
    within the political and economic domains of concrete everyday life.
At each of these spectacular spaces (would-be) hegemonic

representations were made which served
to further buttress the position of those who govern, rule and control
    at a number of social sites simultaneously,
to reinforce or newly inculcate a particular way of seeing,
to make some particular things visible
    and other particular things invisible –
        thereby rearticulating the social and cultural landscape,
to, in short,
    "rearticulate the complex relations among the state,
        the economy and culture."[9]
These very same discourses and representations simultaneously were
aimed
to deflect the counterhegemonic,
to defuse the possibility of cultural and political resistance,
to pacify the potentially disruptive energies of power subjects
    and large op-positioned groups,
to counteract actual or potential dissatisfaction and disenchantment
    with the here and now
        by rewriting history,
        by reinventing collective memory and identity,
        by remythologizing the national past,
        by reenchanting a once-upon-a-time then and there.

The discourses associated with each of these spectacular spaces were would-be hegemonic discourses in the sense that they did not go unanswered. In all three instances the discourses evoked symbolic discontent, or a variety of cultural reworkings – a multitude of struggles over the appropriateness of meanings, over taken-for-granted samenesses and differences, over survival of the custom-ary, over the legitimacy of namings – over and over again, in all of these ways, over the preservation of identity. Some of these cultural contestations progressed beyond off-stage or subterranean expression, becoming translated into spontaneous public actions or organized forms of resistance. However, the great majority of these cultural reworkings did not go beyond symbolic challenge, amounting to little more than an adaptive safety-valve, and thereby contributing to the real-ization of that which was already becoming.[10] All of these cultural reworkings sprung from the complex and contradictory terrain of everyday experiences, from the specific knowledges and awarenesses of incongruity that came with corpo-real engagement in on-the-ground quotidian practices and their associated power relations. *(By the same token, such bodily engagement in the everyday practic(e)al world produced the idea-logics and*

particular knowledges of those people within power-holding fractions who intentionally or unreflectingly promulgated (would-be) hegemonic discourses.)
All of these cultural reworkings, like the spectacular spaces of consumption and hegemonic discourses to which they refer(red), thereby emerged out of and contributed – in smaller or larger ways – to the conditions of Swedish modern(ity).

Every one of Europe's capitalist commodity societies has experienced successive, but differently timed, periods of industrial modernity, high modernity and hypermodernity – periods in which consumption and (would-be) hegemonic discourses were (and remain) intricately enmeshed with one another in distinctive ways, despite being influenced by other parts of Europe. Every capitalist commodity society is currently characterized by a tumultuous, volatile and crisis-strewn hypermodernity, by hypermodernity in juxtaposition with residues of industrial modernity and high modernity, by modernity accentuated and sped up. Every capitalist commodity society has had its spectacular spaces of consumption, its spaces where the spectacle has served (or currently serves) not only as an intervention in the emergence and perpetuation of new forms of consumption, but also as a vehicle for the introductory or continued promotion of (would-be) hegemonic discourses.

In each one of Europe's capitalist commodity societies industrial modernity, high modernity and hypermodernity have unfolded and overlapped with one another under geographically and historically specific conditions, have emerged out of and contributed to the transformation of already existing and nationally distinctive capitalisms,[11] political circumstances and forms of collective consciousness. In each capitalist commodity society industrial modernity, high modernity and hypermodernity have had their place-specific manifestations, their array of (geographical hi)stories in which the local has intersected with the extralocal and the "global," in which new forms of domestic and foreign capital – new forms of production and *consumption* – have intersected with already existing – more or less deeply sedimented – everyday practices, power relations and sets of local and wider taken-for-granted meanings.[12] Unavoidably then, as pivotal spectacular spaces of consumption have appeared in different capitalist commodity societies, they have produced different confrontations between the new and the already existing, have generated different symbolic ruptures, have resulted in singular cultural reworkings of the spectacularly marketed goods and (would-be) hegemonic discourses associated with industrial modernity, high modernity and hypermodernity.

Given that the hypermodern present in Europe is not a homogenized moment, given that it inescapably contains geographically and historically distinctive elements as well as shared characteristics, the interrogation of modernity in Sweden over the past hundred years through the prism of three Stockholm-situated spectacular spaces of consumption arguably takes on broader significance. For where pronounced variety exists around a core terrain of intersections and commonalities, of complex interdependencies and mutual influences past and present, it is the critical examination of extreme cases which can prove most revealing of the central. And, Sweden, after all, has been and continues to be an extreme case. It was not without reason that the country served as an international symbol of modernity from the 1930s through the 1960s.[13] There industrial modernity eventually resulted in the world's highest level of capital concentration in (commodity-promoting) large corporations.[14] There the high modern came to dominate longer than in any other captialist commodity society. There the practices of social welfare and the power relations of social engineering came to permeate daily life more than in any other capitalist commodity society. There, consequently, the dissolution of high modernity into hypermodernity has proved especially unsettling, if not convulsive.

Make no mistake about it. This book is no mere commentary on contemporary Stockholm and Sweden at large. It is written – via its peculiar focus, its minute details, its down-to-earth factuality – as a critical meditation on Europe's various capitalist modernities, as a critical meditation on the precarious hypermodern present and the industrial-modern and high-modern moments which speak to it.

Once again . . .

WALTER BENJAMIN (1982), 574 as quoted in Susan Buck-Morss (1989), 73

Method of this work: literary montage. I have nothing to say, only to show.

SUSAN BUCK-MORSS (1989), 359

The Passagen-Werk *suggests that it makes no sense to divide the era of capitalism into formalist "modernism" and historically eclectic "post-modernism," as those tendencies have been there from the start of industrial culture. The paradoxical dynamics of novelty and repetition simply repeat themselves anew.*

Modernism and postmodernism are not chronological eras but political positions in the century-long struggle between art and technology. If modernism expresses utopian longing by anticipating the reconciliation of social function and aesthetic form, postmodernism acknowledges their nonidentity and keeps fantasies alive. Each position thus represents a partial truth; each will recur "anew" so long as the contradictions of commodity society are not overcome.

[In his One Way Street (Einbahnstrasse) and the notes and drafts for his Arcades Project (Das Passagen-Werk),[15] in his efforts to critically re-present and de-myth-ologize the circulation of commodities] Benjamin effectively "spatialised" time, supplanting the narrative encoding of history through a textual practice that disrupted the historiographic chain in which moments are clipped together like magnets. In practice this required him to reclaim the debris of history from the matrix of [linear, progressive] systematicity in which historiography had embedded it: to blast the fragments from their all-too-familiar, taken-for-granted and, as Benjamin would insist, their mythical context and place them in a new, radically heterogeneous setting in which their integrities would not be fused into one. This practice of montage was derived from the surrealists, of course, who used it to dislocate the boundaries between art and life.[16]

Benjamin sought to bring about an explosion which would bring down the Dream House of History by forcing a discarded, forgotten, even repressed past into an unfamiliar, unreconciled constellation with the present. Like the surrealists, his intention was to shock.

DEREK GREGORY
(1991), 26, 28

As defined by Benjamin, neither montage nor quotation suggests stasis as they now do in relation to structuralist and post-structuralist theory. Rather they contribute to a dialectical reading of history – history defined not as a continuum projected out of the past and propelled by progress into the future, but history apprehended from our vantage-point in the present as ruptured moments that take on significance because of their relationship to the present.

SUSAN WILLIS
(1991), 41

Each of the chapters that follow is in itself a montage, a simultaneous showing of different (geographical hi)stories.

The separate chapters on the Stockholm Exhibition of 1897, the Stockholm Exhibition of 1930, and "The Globe" together constitute a montage of the present, a history of the present in montage form, an assemblage of images. More precisely, they constitute a set of distant, not so distant and very recent (geographical hi)stories, a totality of fragments which brings the past into tension-filled constellation with the present moment, which speaks to the here and now in strikingly unexpected but potentially meaningful and politically charged ways. In ways which are not necessarily confined to here and now Sweden. In ways which may spark a flash of destabilizing re-cognition; a fleeting moment of dis-place-ment, inducing rethinking among readers enmeshed in other late twentieth century societies where the spectacle has become an everyday item of consumption, an unavoidable commonplace wherein (would-be) hegemonic discourses are projected.

And, to the extent that this montage of the present brings industrial modernity and high modernity into confrontation with hypermodernity, to the extent it actually opens up the possibility of a different recognition of how modern politics, economy and culture are inseparably intertwined, it is meant to arrest the attention of the critically inclined, to provoke others into the practice of "heretical empiricism,"[17] to prove suggestive of how contemporary settings – other than Stockholm in particular and Sweden more generally – might be made intelligible, of how other geographically specific configurations of right-now commodity society might be interrogated and re-presented.

Benjamin's "montage" is a highly suggestive point of departure.
A source of intellectual inspiration.
Not a work of art to be mechanically reproduced.

Benjamin's writing resonated with the condition(ing)s of modernity
    in Berlin during his childhood,
    in Berlin and Paris during the twenties and thirties.
Inescapably,
our writings resonate with the condition(ing)s of hypermodernity
    in our particular worlds during the nineties.

Benjamin did not make it over the mountains with his baggage.[18]
We are still carrying ours.
As well as some of his.
Time to unpack!

In attempting to re-present the present
    and its modern antecedents

through the medium of montage,
through the juxtaposing of verbal and visual fragments,
through the juxtaposing of quotations,
> newspaper reports,
> anecdotes,
> song lyrics,
> jokes,
> assorted ethnographic evidence of symbolic discontent,
> aphorisms,
> and more conventional summary statements and notes

with photographs, art reproductions and cartoons,
one may attempt to art-iculate
as well as resonate at several levels at once.

Through assembling (choice) bits
> and (otherwise neglected or discarded) scraps,

through the cut-and-paste practices of montage,
one may attempt to bring alive,
to open the text to multiple ways of knowing
> and multiple sets of meaning,

to allow differently situated voices to be heard,
> to speak to (or past) each other
>> as well as to the contexts from which they emerge
>>> and to which they contribute.

Through deliberately deploying the devices of montage,
one may attempt, simultaneously,
to reveal what is most central to the place and time in question
by confronting the ordinary with the extraordinary,
> the commonplace with the out-of-place,
> the (would-be) hegemonic with the counterhegemonic,
> the ruly with the unruly,
> the power wielders with the subjects of power,
> the margin definers with the marginalized,
> the boundary drawers with the out-of-bounds,
> the norm makers with the "abnormal,"
> the dominating with the dominated.

Through the combinatory possibilities of montage,
one may attempt to uncover what is otherwise covered over,
and, simultaneously,

bring into significant conjuncture
    what otherwise would appear unrelated,
blast into isolated disjuncture
    what otherwise would be buried with insignificance;
over and over again
focusing on tiny, seemingly inconsequential details,
    on fragments of no apparent significance,
    on little brute facts,
so as to project the largest possible picture,
so as to provide an intelligible account –
    an array of truthful knowledges –
which is as partial as possible.

Through the calculated arrangements of montage,
one may present – from re-presented particulars –
    a particular interpretation,
and, simultaneously,
allow theory and theoretical position to speak for themselves,
    to emerge from the spatial (juxta)positions of the text,
        from the silent spaces

    which force discordant fragments
    to whisper and **shout** at each other in polylogue.

Through the de-signing and re-signing designs of montage,
one may confront the reader
    with the possibility of seeing and hearing
        what she would otherwise neither see, nor hear,
    with the possibility of making associations
        that otherwise would go unmade,
by subtly demanding that the meaning of each fragment
    be enhanced and shifted repeatedly
    as a consequence of preceding-fragment echoes
        and subsequent-fragment contents.

In the end,
through all of these simultaneous strivings,
through the maneuvered configurations of montage,
through the intercutting of a set of (geographical hi)stories,
through a strategy of radical heterogeneity,

through (c)rudely juxtaposing the incompatible and contradictory,
one may attempt to bring component fragments
    into mutual illumination,
    and thereby startle.

One may, in other words,
attempt to illuminate,
    by way of shock,
attempt to jolt out of position
    by suggesting a totality of fragments,
attempt to destabilize
    by way of a stunning constellation,
without insisting upon a closure
    that does not exist.

In the process,
    explicitly and implicitly,
raising as many questions
as are answered.

All the while
unavoidably
showing as much about oneself
    as about anything else,
showing as much about one's place in a hypermodern world,
    as about the in-place practices and power relations
        of the hypermodern world itself.

===============

To be for montage
is *not* to be for pastiche –
    for a jumble of atomistic elements,
    for a whimsical hotchpotch,
        to which there is nothing more.

To be for montage
is to be for a totality of fragments,
in which the p(r)o(s)etics of the textual strategy
are the politics of the textual strategy.
Consciously, rather than by unreflected default.

===============

Lev Kuleshov, the avant-garde Soviet film-maker and theoretician, referred to his own particular strategy of montage, his rapid intersplicing of differently situated people and urban landscapes, as "creative geography."[19]

NOTES

1 Foucault (1972), Benjamin (1982).
2 Actually, "Swedish modern" is a term that first came into (ab)usage in the U.S.A. and elsewhere during the late 1920s. See Chapter 2, note 8, and the text thereto.
3 Cf. Rabinow (1989) on French urban planning and modernity.
4 See Pred (1990a) and the literature cited therein.
5 Schorske (1981), xix.
6 Ruth (1984), 2.
7 The various developments and circumstances sketched in this paragraph are spelled out somewhat more in Chapter 3.
8 Grossberg's singular form is converted to the plural in order to emphasize the multiple, dispersed locations of power within the social world of any modern time and place.
9 Grossberg (1992), 247. For the various, not always compatible senses of "hegemony" and "counterhegemony" that derive from Gramsci (1971) see Williams (1977), Laclau and Mouffe (1985), Scott (1985, 1990), Hall (1988), Ryan (1989) and Grossberg (1992). The multifaceted usage of the two terms here results from an intersection between several of these senses and insights gained from my own previous work on the language of situated practices and popular cultural struggles.
10 The coexistence of these different counterhegemonic expressions is to be expected. Because power relations in any given setting take forms which are historically and geographically specific, "hegemonic struggles can have [a number of] radically different forms in different social formations and national contexts" (Grossberg, 1992, 244).
11 On the existence of a multiplicity of capitalisms see Pred and Watts (1992).
12 Cf. Pred (1990b), Pred and Watts (1992), and Massey (1993) on the processual, becoming nature of places; on the place-specific under late twentieth century conditions as a distinctive, persistently altered, mixture of local and wider economic interdependencies, power relations and cultural meanings.
13 Beginning with Marquis Childs' *Sweden: The Middle Way* (1936), and terminating with *Sweden: The Progress Machine* (Jenkins, 1969) and *Sweden: The Prototype of Modern Society* (Thomasson, 1970), a series of widely discussed books helped create and sustain an image within the Anglo-Saxon world of Sweden as the spearhead of social modernity. On the European continent this image gained widespread currency in the years following World War II through the efforts of Willy Brandt, Bruno Kreisky, and other leading Social Democrats who pointed to Sweden as a

model for reconstruction. For a discussion of these and other matters relating to the mythologization of Sweden see Ruth (1984).

14 On repeated occasions during the early 1980s, the combined annual revenues of Sweden's fifty largest business enterprises exceeded the GNP, a condition made possible by the extent and scale of their foreign operations and holdings. Calculations based on scattered issues of *Veckans Affärer*.

15 Benjamin (1979 and 1982).

16 As a conscious art form, montage had its apparent origins in the photomontages of Georg Grosz, Hannah Höch, other Berlin Dadaists, and especially John Heartfield, who, during the late 1920s and 1930s, brought this mode of political collage "to a pitch of polemical ferocity that no artist has since equalled" (Hughes, 1991, 73). Their images "directly cut from the 'reckless everyday psyche' of the press, stuck next to and on top of one another in ways that resembled the laps and dissolves of film editing, ... could combine the grip of a dream with the documentary 'truth' of photography" (Hughes, 1991, 71). Note Buck-Morss (1989, 60–4) on the parallels between Heartfield's work during the 1930s and that of Benjamin. Also see Bürger (1984, 66 ff.) on Benjamin, avant-garde art and allegory.

17 I am indebted to Michael Watts for introducing me to this Pasolinian term. See Pred and Watts (1994).

18 Fleeing Paris in 1940 after it fell to Hitler's troops, Benjamin picked up a U.S. visa in Marseilles and eventually crossed into Spain with some fellow exiles at Port Bou. Confronted by a local official who threatened (for blackmail purposes) to return them in the morning to France and the Gestapo, he took an overdose of morphine. Had he waited until the following day, he actually would have been permitted to proceed to Lisbon (Peter Demetz in his introduction to Benjamin, 1978, xiv–xv).

19 See Levaco (1974).

CHAPTER ONE

SPECTACULAR ARTICULATIONS OF MODERNITY: THE STOCKHOLM EXHIBITION OF 1897

•

World exhibitions were places of pilgrimage to the fetish Commodity.	WALTER BENJAMIN (1976), 165
[The Chicago World's Columbian Exposition of 1893] was a means of mental transportation and conquest.	PHIL PATTON (1993), 38, 44
The illusion of fruitful progress and harmony stood in contrast to change and conflict outside.	
[I]t was not always easy in Paris to tell where the exhibition ended and the world itself began.... Just as the exhibitions were becoming more commercialized, the machinery of commerce was becoming a means of creating an effect of reality indistinguishable from that of the exhibition.	TIMOTHY MITCHELL (1991), 9, 11
One might venture to say that the twentieth century jumped the gun with the exhibition of 1897.	MARTIN A. OHLSSON (1959), 192
Many of us who were young at the turn of the century got their first strong impression of Stockholm as a city, of Swedish society as a living community, during the beautiful exhibition summer of 1897.	YNGVE LARSSON (1967), 15
Sweden's new industrial capacity and claims could hardly be more magnificently demonstrated than there [at the Stockholm Exhibition of 1897]. Neither before nor after has such a large exhibition been arranged in the country.	EVA ERIKSSON (1990), 180

As a conceptual term, "articulation" is charged with two seemingly disparate sets of meanings, one cultural-linguistic and the other physical. To articulate, to make an articulation, is to give voice clearly, is to produce speech sounds and thereby express impressions, is to utter thoughts and feelings intelligbly to others, and – most loosely – is to represent ideas by verbal or other means in a manner that is highly comprehensible, in a manner that the audience finds meaning-filled. To articulate, to create an articulation, is to join by linkage, is to unite by physically connecting, is to bring into interaction elements that are otherwise discrete and separate. Yet, wherever industrial capitalism assumes new forms, wherever new investments are deployed, wherever economic restructuring occurs, wherever new technologies and labor processes are introduced, wherever new goods and services are produced or made available, these two aspects of articulation are virtually certain to become intertangled. New forms of industrial capitalism cannot appear locally without precipitating new physical articulations between the local and the extralocal or "global," without generating new economic linkages between local units and units at other locations, without necessitating the movement of inputs, outputs, money, information or people between local sites and more or less geographically distant sites. At the same time, new forms of capitalism cannot bring locally transformed practices and power relations without bringing some transformation of locally encountered cultural articulations. These new articulations may involve terms and expressions denoting the new itself; denoting newly erected facilities and newly modified local landscapes; denoting new types of technology and equipment; denoting the organization, details and moments of new labor processes; denoting the new social relations of these processes. Or, they may assume some form of symbolic discontent, some form of cultural contestation, some form of struggle over meaning deriving from the experience of new everyday practices and power relations, from the experience of new material circumstances and new rules of the game, from the experience of disjunction and discontinuity, from the experience of the "modern" shockingly displacing the "traditional," from – in a word – the experience of modernity.[1]

During the final decades of the nineteenth century Stockholm underwent perhaps its most dramatic period of social and spatial transformation – of economic, political, civil, and physical change – largely through the appearance of new forms of competitive industrial capitalism already characteristic of many urban centers elsewhere in Western Europe and North America. As the local economy was restructured in fits and starts, manufacturing became increasingly concentrated in large-scale factories, thereby undercutting the relative role of artisanal and small-scale production.[2] As the manufacture of specialized machinery and other goods expanded, as new investments and their attendant employment multipliers encouraged population growth, as levels of demand oscillated upward, the economic articulations of Stockholm with urban units and rural areas elsewhere in Sweden were modified and intensified just as the flow of goods to and from locations elsewhere in the European world economy underwent repeated changes of content and volume.[3] The alterations in everyday life accompanying these transformations evoked new cultural/linguistic articulations among the inhabitants of Stockholm, and especially among its working-class and periodically employed elements. For example, in conjunction with the experience of new everyday practices and power relations there appeared new occupation-specific languages of production (as well as of construction and distribution) that not only referred to new tools and machinery, to new distinctive tasks and labor-task moments. Each of these modes of articulation subsumed a language of discipline-avoidance and retaliatory tactics, a set of terms for labelling and describing the tactics used to undermine new and stricter forms of labor discipline, to continue consuming alcohol on the job, to steal goods or potential income or profit from one's employer. Likewise, the "folk geography," or the popular, shifting language employed for moving about in the physically inconstant and expanding city, not only involved a large assemblage of "improper" terms and phrases that were intermingled with more standard usages in order to facilitate spatial orientation, to designate locations and routes. This set of articulations also subsumed a language of ideological resistance, of resistance to the ideology underlying both the wholesale renaming of streets by the City Council and the various bureaucratic measures newly introduced so as to control street life. Moreover, each of these surface and subsurface languages – as well as those pertaining to consumption and to social reference and address – were crosscut by a politically charged folk humor, by a playful reworking of the ruptures of modernity. Each of these modes of articulation was suffused with malicious metaphors, with a vocabulary of comic irony and irreverence, with expressions of mocking and ridiculing social inversion, with well-aimed projectiles of laughing lewdness and licentiousness, with high levels of symbolic discontent.[4]

The cultural manifestations of modernity that appear alongside the local introduction of successive forms of industrial capitalism are in considerable measure associated with new patterns of consumption, with new ways of interacting with merchandise, with the commodification of elements of everyday life that were previously uncommodified, with the mass marketing of items that either were once exclusive or are without antecedent, with the new (or continued) reign of commodity fetishism. Upon the initial blooming of industrial capitalism, upon consequent increases in disposable income, upon technologically facilitated price decreases, consumption becomes marked by a whirl of quickly passing fads; by goods whose uses are ephemeral and whose biographies are brief; by a flickering succession of new routines – new organizations of time and space – that become quickly deroutinized, quickly dislocated, quickly removed from the daily paths of women and men; by a phantasmagoria of newly awakened wants and desires that are often rapidly translated into needs and requirements; by a relentless parade of fashions, or goods whose newness is "always-again-the-same" – first making a promise, then betraying with disenchantment, first suggesting dream fulfillment, then crushing the dream.[5]

Although there were occasional downward swings, the 1880–1900 period was one in which the average real income of Stockholm's middle- and working-class residents increased, especially from the mid-1890s onwards. After having reached high levels during the 1870s, food prices stayed within a relatively narrow range over the next two decades, or actually wound up at lower levels in the cases of sugar, milk and potatoes; while the income of particular working-class occupations rose significantly over the same period – on the average by over 50 per cent, and in many instances by over 100 per cent.[6] For those fortunate enough to hold regular full-time employment, such income gains usually meant an increase in the amount of disposable income at their command after expenditure for the basic necessities of food and housing.[7] Thus, as youthful industrial capitalism provided an ever-changing array of alternatives, portions of the working class were, at least periodically, able to supplant old consumption practices with novel ones; to succumb to the latest "tastes;" to buy a new brand of cigarette or snuff; to acquire a currently fashionable piece of apparel or cheap jewelry; to purchase the latest watch model and wear it as a weekend ornament (when on one's own time); to obtain a few articles of newly designed, low-cost chinaware; to buy or borrow a weekly installment of *Lätt på foten* (*Light on the Foot*), *Vackra Marie* (*Beautiful Marie*), *Bakom fällda rullgardiner* (*Behind Lowered Window-Shades*), or any of a great number of cheap-pulp *feuilletons* belonging to the romance, damsel-in-distress and adventure genres.[8]

Swept up in the new vortex of consumption, the well-to-do middle class chose to crowd their apartments with a hotchpotch of bric-à-brac, satin draperies, velvet-covered furniture and other momentarily fashionable items as they simultaneously sought to create an idyllic haven in which to cushion themselves from the threatening and transitory outside world and a showcase of achieved position and status.[9]

"... [E]veryday objects [of consumption] continually assert their presence as simultaneously material force and symbol,"[10] as simultaneously physical thing and meaning.

Culture is, among other things, one with the meanings through which women and men "handle and respond to the conditions of [everyday-practical] existence,"[11] one with the meanings and values used by distinct groups and classes to navigate the flow of everyday life under historically and geographically specific circumstances. Therefore, the in-place transformation of consumption patterns is always one with processes of cultural transformation, one with the birthing and deathing of meanings. Or, the on-the-ground adoption of new and pronouncedly different patterns of consumption involves the ascription and internalization of new meanings, the incorporation of new meanings by which to live and negotiate the everyday. Often these meanings are apt to be built on more than a passive acceptance of what messages the market has to offer, of what representations market interests promote or of what representations promoters attempt to make interest arresting. Often these meanings are apt to have complex origins; are apt to involve an element of active cultural production; are apt to be mediated by previous experiences, dispositions and elements of identity; are apt to be derived from an intermingling – if not a collision with – previously developed knowledge and meanings,[12] from the summoning up and reworking of individual and collective memories.[13] Being constituted in this manner, the cultural transformation that emerges with new and radically different patterns of consumption is apt to engender confusion or discontent to the extent that commercially promoted or actively produced meanings are contradictory to already held meanings, are incompatible with meanings that are central to already existing gender, class, generational, regional or ethnic identities.[14]

The pronouncedly different patterns of consumption that began to take shape in *fin de siècle* Stockholm were adopted by a heterogeneous and sharply differenc-iated population, were given meanings by a population of diverse geographic origins and marked class divisions. Towards the end of the 1890s, approximately 60 per cent of Stockholm's population had been born outside of the city, born for the most part in agricultural or non-urban settings, born in areas that stretched from the southernmost to the northernmost provinces of Sweden,[15] born in regions with distinctive dialects and cultures of their own, born in regions whose then articulations and meanings of everyday life often did not make sense elsewhere.[16] Whether of local or non-local origin, members of the rapidly growing working class found themselves subject not only to new forms of domination at the workplace; not only to political domination by a City Council that solely represented business and wealthy bourgeois interests (a *man's* right to vote, and his asssignment of anywhere from one to a hundred votes, was a matter determined by the size of his property holdings and annual income). They were simultaneously subject to attempts by the bourgeoisie to establish cultural/ideological domination; to bourgeois efforts – in the schools, press and elsewhere – to make their idea-logics and ideal-logics stick; to a wishful bourgeois mythologizing of a simple, harmonious, frictionless past where every woman and man knew their place – and was content with it; to a bourgeois missionizing that stressed Oscarian (or Victorian)[17] morals, that projected a certain world-view of appropriate behavior in private and public spaces, that sought to norm-alize everyday life in the home and on the streets, that insisted upon order, upon everything in its place.[18]

Given these heterogeneous origins and tension-ridden class circumstances, what meanings were act-ually given to new patterns of consumption by the residents of Stockholm?

What articulations resulted among people whose identities were centered on rather different biographical experiences and thereby on rather different meanings?

In major European, North American and Australian urban centers where industrial capitalism began to flourish at some point during the nineteenth century, exposure to its attendant forms of consumption, to the possibilities of new commodities, to new messages as to what money might buy, was not necessarily confined to the quotidian, to ordinary times and spaces, to chance daily-path encounters with shop-window displays, newspaper advertisements, street posters or possessions already purchased by others. Occasionally, the universe of new consumption possibilities was given spectacular articulation, was presented in concentrated form, was displayed in an extraordinary time and space, was exposed to the senses in a time and space isolated from the routines and project demands of the everyday, was put on exhibition to be gazed at under circumstances free from everyday responsibility, was ambitiously and grandiosely promoted at a "World's Fair" or "Industrial Exhibition." As a public space designed to manufacture private desires; as a space suggesting an unlimited profusion of commodities; as a space where the commercial, the political and the cultural were ideologically melted together; as a space intensely projecting dominant meanings of modernity; as a space in which symbols of power were given condensed expression for manipulative or seductive purposes; as a space for the representation of grand claims;[19] as a space of illusory and fantastic images, the space of such exhibitions was a precursor to the "society of the spectacle," to the incessant "spectacle" of merchandise marketing via television and other forms of communication encountered in private as well as public spaces.[20]

The truly spectacular spectacle of the international exhibition – "the ultimate spectacle of an ordered totality" – made its debut in 1851 at London's famed Great Exhibition of the Works of Industry of All Nations. There, at the Crystal Palace, "an ensemble of disciplines and techniques of display that had been developed within the previous histories of museums, panoramas, Mechanic's Institute exhibitions, art galleries, and arcades" were brought together and reordered so as to establish "a new pedagogic relation between state and people," so as to morally and culturally regulate, so as to exercise power and bring order – an assent to governance – by way of conveying messages, by allowing "people, *en masse*" to see rather than be surveilled, "to become the subjects rather than the objects of knowledge."[21] By the early years of the next century mammoth international exhibitions had been held in Paris (1855, 1867, 1878, 1889, 1900), Vienna (1873), Philadelphia (1876), Chicago (1893), Buffalo (1901), Glasgow (1901) and St Louis (1904); while somewhat more modestly dimensioned, but still large-scale, expositions had been arranged in dozens of other cities.[22] Each of "these expositions were upper-class creations initiated and controlled by locally or nationally prominent elites."[23] Each of them served simultaneously as an economic and political battleground; as "festivals of high machine-age capitalism,"[24] as a site where nations amassed their symbols, fought for markets, showed off their industrial prowess, "peddled their self-images"[25] and competed for international prestige. Each of them was a minutely planned environment of representations, an environment in which every (imageinable) thing had its place, in which new patterns of consumption were extolled, in which an often strange new world of goods was represented as everybody's new just-around-the-corner world of everyday life. Each of them was an elaborately contrived setting in which the "observing subject" was presented with a "system of signification" arranged so as to ensure awe and curiosity while simultaneously creating an appearance of order, in which the world was divided into two realms – "the realm of things and the separate realm of their meaning or truth," in which things were set up so as to evoke "some larger truth" regarding either progress and history or "human industry and empire,"[26] in which the public was to learn its lessons by looking. ("[T]he principle of learning by looking was strongly endorsed by contemporary pedagogical theory, which postulated that people learned first of all by observation of everyday objects, secondly by comparison and classification, and only finally through abstract reasoning."[27])

> [A]nd since exhibitions [elsewhere] have proved extremely attractive to [exhibiting] producers as well as the purchasing public, help [in promoting the foreign and domestic sales of Swedish manufacturers] can be delivered in no better way than by the arrangement of such of an event.
>
> From the standing committee report regarding State support to the proposed Stockholm Exhibition (*Riksdagstryck*, 1895d, 6–7).

The Stockholm Art and Industrial Exhibition of 1897 had been in the making since 1893, when some 2,700 industrialists petitioned the State for an international exposition that would include the participation of neighboring countries. Earlier initiatives had been rejected outright by Parliament (1885) or put aside (1890) owing to the World's Columbian Exposition that was to be held in Chicago and in which Sweden was to officially participate during the identical period suggested. Although some politicians questioned the means by which it was to be financed, the proposal for 1897 proved appealing not only because no such event had been arranged in Stockholm since the very modestly scaled exhibition of 1866, not only because a strong upswing in the business cycle provided a new climate of optimism, not only because it was expected to further foster development of the country's manufacturing, but because it was also difficult to dismiss an exhibition that would coincide with the twenty-fifth anniversary of King Oscar II's accession to the throne, a jubilee of central symbolic significance to Sweden's high bourgeoisie.[28] In keeping with the bureaucratic rationality of the time, as well as the spread of specialized-division-of-labor principles, the hierarchical administrative apparatus established in 1895 for the organization of the exhibition was enormous: at the top, a fifty-member Central Committee chaired by Prince Eugén; directly below it an administrative committee headed by the Exhibition's General Commissioner, Baron Gustaf Tamm, the Governor of Stockholm; beneath it separate committees to deal with financial matters, buildings, the press and publicity, transportation, lodging, prize-jury selection and the maintenance of order; also under its jurisdiction – in addition to local committees for Göteborg, Malmö and each of the country's administrative provinces – were an arts division with five separate working committees, and an industrial division, subdivided into eight working committees which in turn had 24 first-level and 105 second-level working committees below them. And, at every level of this hierarchy, from the Central Committee down, those numerically dominating committee membership were industrial entrepreneurs, wholesalers, engineers, factory managers and directors; those who stood to benefit most from the educational objectives of the exhibition; those who had "hegemonic ambitions."[29]

BURTON BENEDICT (1983), 2	*People [attending world's fairs] were to be educated about what to buy, but more basically they were to be taught to want more things, better quality things and quite new things.*
GRAEME DAVISON (1982), 19	*[T]he business of the exhibition was not simply to sell things but to symbolize, and thus to disseminate, the ruling ideals of an industrial age.*
TIMOTHY MITCHELL (1991), 162	*To submit and become a citizen of such an exhibitional world was to become a consumer of commodities and meanings.*
STUART EWEN (1988), 204	*[The international exhibition] was an attempt to erect a self-consciously ideological tableau, a theater of modern power.*

The educational exposure of international exhibition visitors to the object-ive possibilities of new forms of consumption was one with their exposure to representations that glorified domestic industrial achievements, that encouraged national pride, that sought to contribute to the further cultural construction of an "imagined community,"[30] of a national identity based on shared symbols and sentiments, of a nationally oriented collective memory. At the same time, such exposure was one with an exposure to ideological articulations that linked industrial technology with continuous progress and a belief in an ever-better future, to ideological articulations that silently denied the widespread existence of domestic tensions and poverty, and thereby legitimated the present. Thus, the rhetorics of nationalism and progress were articulated with one another as well as with those of consumption.[31]

ELLEN KEY, prominent author and intellectual, writing enrapturedly – if not chauvinistically – in an open letter to Verner von Heidenstam in 1897 (Björck, 1947, 360), while teaching for the Workman's Institute (*Arbetarinstitut*), a state-supported organization whose popular scientific public lectures were aimed at those who had no formal education during their adolescent years.

[The national orientation of the Stockholm Exhibition's art and architecture may well have gone unnoticed by the majority of attendees][32] *but the multitude from the outset clearly perceived what enormous sums of national energy and national intelligence were represented in the areas of large-scale industry and invention, where results had been reached that unambiguously show that not only a large part of the nation's bold eagerness to accomplish, but also its creative imagination has taken off in this direction during recent decades. And not one among all the hundred-thousands who surged through the exhibition was completely unmoved by the sense of national solidarity. Everyone understood that this was common property, a common pride; everyone met together in a happy awareness: My people have been capable of this; these glories of labor are mine; so great are the resources of my country; so rich is*

the nature of my country; so strong and so gifted is my nation. . . . Everyone met together in a great [outburst of] national triumphant joy which brought people of different regional and class backgrounds closer to one another.

Many exhibitors – and this is especially valid for quite a number of large-scale industrial firms which are exhibiting in their own pavilions – have realized that both their own and national honor demands that they take great pains to demonstrate (good) taste and judgment in the arrangement of their displays.

Dagens Nyheter, May 25, 1897

"It is . . . a pleasure to be able to note the unanimous recognition which has come to the exhibition from every direction, from home and abroad, and from truly competent authorities."[33] . . . Numerous articulations from observers near and far, articulations disseminated in the daily press throughout the May 15–October 3 duration of the exhibition, articulations confirming the success of the spectacle, articulations repeatedly testifying to the enterprising spirit and genius of Swedish industry, articulations buttressing the nationalistic sentiments promulgated by the spectacle. . . . "Stockholmers are not just a little proud about the exhibition, and their desire to know the opinion of foreigners reminds me of the Americans' never-ending 'What do you think about the United States?' They can calm down: the Stockholm Exhibition is unmatched among its kind."[34] . . . "The Stockholm Exhibition is better than last year's in Berlin and all the others I have visited save Paris in terms of its overall harmonious effect. . . . [With respect to the special exhibits] the Swedes possess a quite special ability – greatly surpassing both the Germans, Danes and Norwegians – to arrange something that is at one and the same time stylish, practical and comprehensible."[35] . . . "In many respects it surpasses all previous exhibits I have seen. . . . Its site is captivating – surpassed only by that of the Chicago World's Fair."[36] . . . "[The contours and appearance of] the Hall of Industry are impressive, fantastic and attractive. . . . The Swedes are great inventors and excellent engineers and mechanics, to which the numerous displays [of the Machinery Hall] bear witness. . . . *Djurgården* [the site of the exhibition] is one of the most beautiful places in the world."[37]

[The international exposition is] an educational institution where the exhibitors are teachers, the exhibited objects are teaching materials, and the visitors are pupils.

S.A. ANDRÉE
(1897, in a second edition reissued shortly before the Stockholm Exhibition, as quoted by Ekström 1991a, 4)

[The public attending the Stockholm Exhibition] will all the more realize that many articles previously fetched from abroad can be obtained – without

Riksdagstryck (1895a), 15

loss of quality and just as cheaply – within the country's own borders.

Object lessons were to be provided not only to public service employees, professionals, property owners, shopkeepers and members of the middle class in general; but also to members of the working class, to those who often lived in crowded conditions – frequently "without access to light and fresh air"[38] – and were subject to unemployment without warning; to those who were generally without the means to obtain much of what was displayed before them; to those whose organizations now rumbled for social justice; to those whose increasingly loud political movements were a source of bourgeois anxiety; to those who inventively resisted in the hidden nooks and crannies of everyday life; to those who were seen as uncivilized, undisciplined and lacking "culture," and therefore already being made subject to other forms of reforming knowledge.

Nordens Expositionstidning, January, 1897[39]

We have with great pleasure seen ... decisions undertaken by cities and industrial companies to allocate funds for subsidizing the travel of industrial workers who wish, for the benefit of themselves and others, to visit the Exhibition and acquire learning from its diverse educational materials.

Repeatedly, side by side – in the Machinery Hall and the Hall of Industry, in other "halls" and in the pavilions of individual firms – were the latest goods or elements of production technology and their outmoded predecessors.[40] Here there was no ambiguity, here was a bald articulation of modernity: machinery and finished products as material signifiers of progress.

Hear the whistling steam,
it boils and seethes!
The machinery grates
with roaring tempo.
See how harnessed energy has broken
a shortcut to its goal,
new implements cast
from a supple steel!
How she ingeniously has fashioned
everything for the demands of life![41]

Four hundred voices, performing part of a specially commissioned cantata at the Stockholm Exhibition's opening ceremonies, chorally praising the achievements of industrial technology, voicing the "economic progress made in almost every branch of industry since the preceding 1866 exhibition,"[42] pronouncing a progress that meets all human demands, legitimizing lyrics by Count Carl Snoilsky, member of the Swedish Academy.

Wishing away what was currently disturbing, defusing any suggestion of domination, opening the window to a future even more filled with progress, sentiments such as the following were declared both during and after the exposition: "We believe that the exhibition ... has taught us that Sweden's people have within their grasp – by way of continued united endeavor and perseverance – to lift their country to ever higher levels of development in all respects."[43]

Nearly every international exhibition began with a procession in which the officials of the host nation and visiting dignitaries advanced through the city to the site of the exhibition itself. The spectacle of the assembled nations proceeding in step towards the palace of art and industry fittingly symbolized those ideals of linear progress and international concord that inspired the exhibition movement, and confirmed the social order of the host nation.

GRAEME DAVISON
(1982), 10

In the mid-May beginning, a State-ly pageantry whose culminating ceremonies are reserved largely for the selected few, for the high-hatted and socially high. On this day, at least, more or less total segregaton of the uncommon few and the common many. On this day, working- and middle-class throngs along the flag-forested Strandvägen route to the exhibition site, awaiting glimpses of splendor. On this day, name-day Sofia, name day of the Queen, a crowd of the chosen assembled at noon before the Hall of Industry – the exhibition's principal architectural icon – a stunningly fantastic building with a quartet of minarets and a central cupola inspired by Istanbul's Hagia Sofia mosque. In attendance, the Royal Family and staff, the Crown Prince and Princess of Denmark with children, the Swedish and Norwegian cabinets, other dignitaries from Norway, Denmark and Russia (the other nations represented by pavilions or extensive industrial displays of their own), the entire diplomatic corps colorfully preening as birds of paradise, members of parliament with their wives, the Stockholm City Council, functionaries associated with the exhibition's various administrative committees, rifle-bearing troops of the Svea Guards decoratively attired in plumed helmets and parade dress, and, here and there, exhibition ushers and watchmen in their light-blue uniforms. In all, 1,500 invited guests and 654 individuals able to afford the ten-crown cost of an opening-ceremony ticket. (In 1897, ten crowns would have been the equivalent of about 2.5 days' wages for a dockworker or other heavy laborer.)[44]

Following the drum-rolls, parade music, base-toned shouts of miltary command, and distant sound of cannon salutes that accompanied the arrival of the royal entourage; following brief words of introduction by the Crown Prince, by Prince Eugén, and by General Commissioner-Governor-Baron Tamm; following a singing of "From the Depth of the Swedish Heart" by a now-standing, bare-headed audience; following the jubilant chorus and opera star performance of Count Snoilsky's cantata; a booming-voice speech by the King. Majestic rhetoric, florid similes and images of contrast: the ongoing Greco-Turkish war in proximity of the Hagia Sofia mosque contrasted with the peaceful (commercial) competition about to unfold within and around the building before which he stands; an invocation of Thule, "the ancient fairyland" of northern climes; nationalism-promoting articulations of the highest order tempered by friendly references to our "brothers" in Norway (still bound to Sweden by political union), our "kinsmen" in Denmark, our "neighbors" in Russia. Immediately afterwards, in rapid succession: four resounding hurrahs in response to a "Long live the King" from the Crown Prince; blaring trumpets from a minaret-to-cupola flying bridge above; the firing of cannons; the pealing of church bells; a momentary burst of sunlight from among the clouds.[45]

> That exhibition rhetoric, with its stress upon ideals of economic progress and social harmony, was so dramatically at odds with contemporary reality is not perhaps to be wondered at, for, as anthropologists would tell us, the very purpose of such rituals as the opening of the Great Exhibition was to offer a symbolic resolution of the contradictions of social life.

GRAEME DAVISON (1982), 10

Sofia Day press: "As work opportunities for construction woodworkers in Stockholm will with all certainty diminish significantly during the near future, owing to the completion of projects currently connected with the exhibition, the steering committee of the Construction Woodworkers Union warns union members and all other woodworkers throughout the country not to travel to Stockholm in order to seek employment without first, in each and every case, accurately informing themselves of a position there before leaving their current job."[46]

The Stockholm press of some eleven weeks earlier: brief strikes at several of the exhibition's construction sites reported.[47]

> [T]he clear relations between the new technology and [French] imperialism, between the modes of production and the expansion of markets, were also spelled out in these [Parisian] exhibitions [of the late nineteenth century].

DEBORA SILVERMAN (1977), 70

The parallels between the Stockholm Exhibition of 1897 and exhibitions earlier held in Paris, Chicago, Berlin, London and elsewhere; the conscious and subconscious elements of imitation evident in the form and content of the Stockholm Exhibition; were themselves in some measure a manifestation of the city's articulations with the European world economy.

Unstopped for the day, or for the duration of the exhibition, were the pulsating economic articulations of Stockholm, were the pulsations which found new symbolic articulations in the spectacle of the exhibition itself, were the economic pulsations that reverberated with the spectacular consumer-education and national-identity construction projects occurring at Djurgården, were the goods flows to and from the city, were the import–export and domestic trade articulations demonstrating that Sweden's political capital was also the capital of Swedish capital.

Among the goods examined by Stockholm customs officials during 1897: 413,621 metric tons of coal, largely from England, but also from Germany, France and the U.S.A.; 11,042 tons of machinery, machinery parts and tools from a similar set of origins; 8,183 tons of coffee and 16,225 tons of sugar from various tropical locations.[48] In terms of shipping tonnage, England, Russia (including Finland) and Germany respectively accounted for approximately 41, 26 and 14 per cent of all foreign arrivals and departures at about this time.[49] Among the most important foreign-bound goods in terms of value were the various types of specialized machinery, tools and equipment shipped to Finland and Russia.[50] While Stockholm manufacturers were providing printed matter, china and porcelain for the country at large; while they turned out agricultural and dairy machinery for central Sweden; while they sent new types of sawmill machinery to the Bothnian ports of *Norrland* and steam engines and recently developed producers' goods to more widespread small-town industrial facilities; large quantities of domestic goods were simultaneously brought into the city by rail and coastal and inland shipping vessels. Lumber from *Norrland* and more than 35 million bricks, mostly from around Lake Mälar,[51] arrived for construction purposes. Iron and steel from works in central Sweden arrived to feed the city's burgeoning manufacturing sector, while the city's expanding population was fed by grains and other foodstuffs that arrived principally from its immediate hinterland and *Skåne* (as well as from abroad).

SUSAN BUCK-MORSS (1989), 253

The visible theoretical armature of [Walter Benjamin's] Passagen-Werk *is a secular, sociopsychological theory of modernity, of modernity as a dreamworld.*

SUSAN BUCK-MORSS (1989), 359

The Passagen-Werk *suggests that it makes no sense to divide the era of capitalism into formalist "modernism" and historically eclectic "post-modernism," as those tendencies have been there from the start of industrial culture.*

ROSALIND WILLIAMS (1982), 12

The expositions ... displayed a novel and crucial juxtaposition of imagination and merchandise, of dreams and commerce, of collective consciousness and economic fact. In mass consumption the needs of the imagination play as large a role as those of the body. Both are exploited by commerce, which appeals to consumers by inviting them into a fantasy world of pleasure, comfort, and amusement.

GRAEME DAVISON (1982), 8

The transformation of the exhibition from a simple display of technology into a national, and quasi-religious, festival was manifest in the changing forms and confused vocabulary of its architecture.

[According to Simmel's view on the aesthetics of industrial exhibitions] the fleeting life of the commodity is also reflected in their architecture. Thus, whereas the architecture reflects "the conscious negation of the monumental style," "the character of a creation for transitoriness" becomes the dominant impression. This transitory impression must still embody some of the "eternity of forms" in order not to totally reveal the illusory nature of the seemingly permanent character of the contents of such exhibitions.

DAVID FRISBY
(1986), 95; quotes from
GEORG SIMMEL
(1896)

[A]rchitectural representation is never pure.... [It] is always colored by power relations.

ZEYNEP CELIK
(1992), 195

It was an improvised architecture which didn't remind one of Sweden, but more of the Orient.

AUGUST STRINDBERG
Götiska rummen, as quoted in
Ulf Sörenson (1991), 140

To approach the Stockholm Exhibition by bridge or by an airborne, cable-suspended gondola; to cross over to what had been *Lejonslätten* (The Lion's Plain), to a piece of land whose green grass had for decades proved attractive to working-class Sunday strollers and picnickers;[52] to complete that rite of passage from everyday spaces to the space of the spectacular; was to enter a fantasy world, a world of dreams, a world of illusions. Here, in the terms of those who directly experienced it, was "a fantasy city," "a summer fairy-tale," "a toy city," "a dazzling painting, a sparkling jewel from *A Thousand and One Nights*."[53] Here was a landscape whose Disneyland-like architectural rhetoric displaced the popular clamor for political (voting-right) freedom with images regarding the freedom to consume. Here was a landscape whose architectural forms suggested a well-decked *smörgåsbord*, just waiting to be consumed.[54] Here was a built landscape whose "uncommonly successful" physiognomy gave "an extraordinarily picturesque and pleasing architectural impression."[55] Here was supposedly-before-its-time "postmodernism," a pastiche of architectural styles, a Babel of architectural expressions, an architectural wedding of the exotic and the national, the "oriental" and the Nordic. Here was the Hall of Industry, a queenly centerpiece blithely appropriating from a religious centerpiece of Islamic architecture,[56] derived from the Hagia Sofia and yet with a cross atop each of its fifty-meter-high minarets, a tremendous tiara atop its central cupola, and that representation of the Queen's headpiece itself topped by a lengthy flagpole.[57] Here was the Hall of Arts which – by way of

its white, windowless entrance façade and frieze-topped walls – suggested the Spanish and yet pointed toward the "more austere, organized simplicity" of newly emerging Swedish modernist architecture.[58] Here was the Forest (Industry) Pavilion, dominated both by tree poles clustered into absurdly phallic towers beside a completely windowed façade, and by the use of twigs, pinecones and spruce-bark chips in elaborately decorative devices on its triangular veranda as well as on its main structure – a joining of the naturalistic and the Jugend which Ellen Key regarded as a "brilliant" reworking of traditional Swedish motifs, an expression of true patriotism triumphant.[59] Here there were towers seemingly everywhere; phallic representations of the patron-izing, patriarchal power of new factory owners; representations sometimes remarkable for their undisguised anatomical accuracy. Here was the reconstruction of "Old Stockholm," with its spires and towers, its dense orgy of renaissance forms; the Stockholm Pavilion, with its sculptured ornamentation and general features reminiscent of the baroque, of eighteenth-century high Swedish architecture; and so on. Here was a scattering of pavilions and buildings that insisted upon an unambiguous

unification of physical form and product promotion, of archtiectural contours and larger-than-life messages of choice construction. Here was *"architecture parlante"*: the pavilion of Liljeholmen's Candle Wax Factory erected in the form of a candle so gigantic that it served as a landmark of orientation for many;[60] the Gustaf Piehl's Brewery Pavilion partly in the shape of a beer barrel and the St Erik's Brewery Pavilion almost entirely in that shape; the Sandviken Ironworks building with its gateway in the form of a huge steam hammer; the Bofors (munitions) Works Pavilion in the guise of "a mighty armored turret, from which two colossal cannons stick out, totally fear inspiring despite being made of nothing but wood;"[61] the Finspång's (munition works) Pavilion, with its twin towers in the shape of enormous artillery shells; and Löfberg & Co., producers of what was then the national alcoholic drink of the bourgeoisie, represented in the middle of a lawn by a stupendous bottle of Monopol *punsch* (arrack) atop an ornate pedestal. And, here was another structure that conjoined architectural form with merchandise marketing, but with an added dimension of glass-clear gender ideology: The Lady's Pavilion, sponsored by Herm. Meeths Co. (a chain of stores retailing sewing materials) was fashioned in the form of an elegant sewing box. This round symbol of domesticity was reserved for the use of women who wished a respite from the exhibition in a womb-like environment of silk-draped walls and soft sofas.[62]

After 1851, world fairs were to function less as vehicles for the technical education of the working classes than as instruments for their stupefaction before the reified products of their own labor, "places of pilgrimage," as Benjamin put it, "to the fetish Commodity."

TONY BENNETT (1988), 94

The close proximity within which the most heterogeneous industrial products are confined (at international exhibitions) produces a paralysis in the capacity for perception, a true hypnosis.... in its fragmentation of weak impressions there remains in the memory the notion that one should be amused here.

GEORG SIMMEL (1896), as quoted in David Frisby (1986), 94

The word ["objective," which first came into vogue in the middle of the nineteenth century,] denoted the modern sense of the detachment, both physical and conceptual, of the self from the object world – the detachment epitomised, as I have been suggesting, in the visitor to an [international] exhibition. At the same time, the word suggested a passive curiosity of the kind the organisers of exhibitions hoped to evoke in those who visited them.

TIMOTHY MITCHELL (1991), 20

The international fairs were the origins of the "pleasure industry" which "refined and multiplied the varieties of reactive behavior of the masses. It thereby prepares the masses for adapting to advertisements, the connection between the advertising industry and world expositions is thus well-founded." At the fairs the crowds were conditioned to the principle of advertisements: "Look but don't touch," and taught to derive pleasure from the spectacle alone.

SUSAN BUCK-MORSS (1989), 85, quoting WALTER BENJAMIN (1972–89), V, 267

An English visitor ... noted ... that the fair [the World's Columbian Exposition held in Chicago during 1893] resembled the contents of a great dry goods store mixed up with the contents of museums.

WILLIAM CRONON (1991), 344[63]

Enter into the Hall of Industry, enter into the Chemico-Technical Industry Building, enter into the pavilion of Scanian Large-Scale Industry, enter into any one of a number of lesser halls or pavilions where the products of Swedish capital are spatially segregated from those of Norwegian, Danish and Russian capital. Enter into the interior of the exhibitionary dreamworld, the spectacular dreamworld of the commodity form, the phantasmagorical world

of disconnected consumer-object images. Join the anonymous throng moving from fragmented display to fragmented display. Be swept along in "the sea of people."[64] Spin around, be spun around, among the tightly packed, costly framed displays of 6,781 exhibitors.[65] Move by new products never seen before, old commodities newly fashioned, new and old commodities matter-of-factly arranged for simultaneous enlightenment and seduction, products dreamily displayed – adornments placed in a scene – so as to stand out, so as to appear more remark-able than those around them. Move by the mass-produced and the custom-made, the affordable and the luxurious, the within reach, the just beyond the grasp and the totally out of reach. Move by messages making promises of things to come, straightforward messages, yet messages visually whispering in shifting suggestive tones: "You can have me now;" "Come hither, as soon as you can, I'll be waiting for you." See the already familiar made desirous. See beers, cigarettes, cigars, snuffs, corsets and other underclothes, bird cages, knives, locks, textiles, margerines, crackers, coffees, shoes, stationeries, umbrellas, hats, toys, soaps, marmelades, chocolates, preserves, scissors, razors, stoves, flours, potato starches, kitchen utensils, ribbons, paints, varnishes, towels, candles, matches. See the modern of the moment and the modern of the future. See light bulbs, telephones (at least fourteen different models from L.M. Eriksson), streamlined bicycles (no more the velocipede with its exaggerated back wheel), calculators, can-openers, photographic equipment, sleek sailing boats, vacation or racing skis and skiing attire, new sewing-machine models, rubber pipes, washing machines, mechanized clothes wringers, steam pressure-cookers, refrigerators, water-proofing liquids. (Outside, in the open-air dreamworld, one might see Stockholm's first automobile, a personally owned vehicle capable of doing twenty-five kilometers per hour.) See moral instructions, see homely bourgeois reminders that a private sphere – a secluded social island – of comfort, warmth and coziness is essential to the good and orderly life, to a cultivated existence.[66] See rugs, curtains and draperies, pianos, urns, vases, candelabra and other decorative objects in gold or silver, crystal chandeliers, wall clocks, glassware, usually soft and more or less ornate furniture of different styles, figurines of marble or china, elaborate lamps, wallpapers of intricate design, entire living-room ensembles. Absorb one disjointed impression after another. Be relentlessly bombarded by disjointed stimuli. Be cumulatively shocked into a dreamy state of passive enjoyment. Momentarily gaze at display after display where the commodity is shown in profusion rather than isolation, where the eyes are confronted by fifteen bicycle models rather than one, by dozens of shoes or hats rather than a few, by hundreds of cans of fish or steel

kettles and pans rather than a small number. Have shared imprints made upon your consciousness. Allow the construction of your memory to become a part of the construction of a then-and-there collective memory. Absorb images of product promotion that make common sense by merging form and content. See the display of the Vestervik Match Factory (*Vesterviks tändsticksfabrik*) entirely covered with matchbox labels. See the display of the Scandinavian Pen Factory (*Skandinaviska pennfabriken*) – "we sold 16 million in Sweden and abroad last year" – focused on one monstrous pen in gleaming steel. See the display of the Bolt Factory Corporation (*Bultfabrikensaktiebolaget*) in the form of an enormous bolt covered by countless nuts, bolts, nails and rivets of various size. See the display of the Scandinavian Benedictine Co. (*Skandinaviska Benedictine-kompaniet*) – a pyramid of bottles housed within a hermit-monk's small hut. See the display of the Central Printing Works Stamp Factory (*Centraltryckeriets stämpelfabrik*) in the form of a gigantic stamping device. See numerous other displays where what is most prominent is the exhibiting company's trademark in overblown format, where what is most prominent is that mark of distinction, that mark setting a company's products apart from those of its competitors, that mark guaranteeing that one could always get the same supposedly superior good, that mark shaping an *image*. See, for example, the display of Johansson & Carlander, a cotton-thread manufacturer, flanked by two imitation-bronze statues of Hercules leading a struggling lion by a cotton-thread leash.[67]

Line from a fictional letter in a popular humor magazine, *Söndags-Nisse*, June 6, 1897

But I don't know what I saw, because I saw so much.

GRAEME DAVISON (1982), 7–8

The focus of most nineteenth-century exhibitions was the machinery hall – usually the largest and most central building in the show – where hundreds of machines thumped and whirred to the rhythm of a great steam-engine. It was here, above all, that inanimate nature most clearly assumed the human qualities of life and movement.

CYRUS WALLÉN (1971), 183

What wonders were not to be seen there! ... a colossal steam-engine with upright cylinder. It was in operation, extremely slowly [when I was there as a 12-year old]. One of the tailors claimed that they didn't dare let it go at full speed, or even at half speed, because otherwise such a hefty guy would certainly shake the hall to bits.

THE STOCKHOLM EXHIBITION OF 1897

I can't forget the big enormous Machinery Hall, where one saw lots of ingenious machines, and some of these machines could be seen in full operation . . . one understood how these machines simplified work and made everything easier for people. I especially remember an envelope-producing machine that caught my attention. One got to follow its workings and see how the manufacturing was done from the initial placement of paper in the machine; one saw how it cut the paper, spread on paste and folded, in the end issuing a finished envelope. It was very interesting to see how fast and dexterously the machine worked. Such work had previously occurred mostly by hand and demanded a lot of time.

AXEL ADOLF OLSSON
(1947), 1–2

Enter into another phantasmagorical interior. Enter into the interior dreamworld of the Machinery Hall or the pavilions of firms such as E. Hirsch & Co., de Lavals and the Separator Corporation. Enter into the dreamworld of producers' goods, into the dreamworld without which there would be no dreamworld of the commodity form, into the dreamworld of spectacular national accomplishment. Go from one awe-inspiring, pacifying lesson to

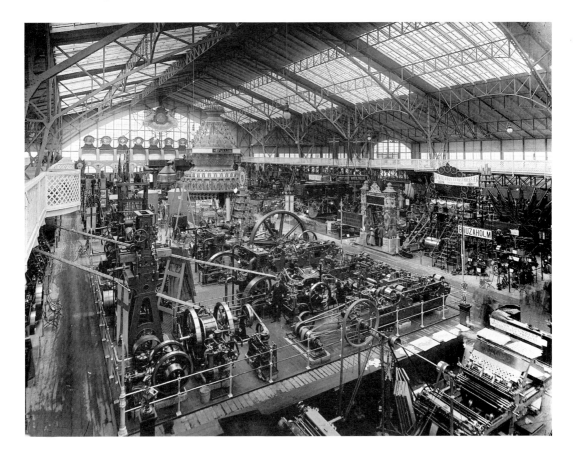

another. Be dwarfed. Gaze with the ear, as well as the eye, until transported. Be stunned and enthralled by machines which "promise ... the naturalization of humanity and the humanization of nature [but sooner or later] result instead in the mechanization of both."[68] Gape and gape again at steam turbines, ship engines, textile-spinning and weaving machinery, dynamos, transformers, dairy machinery, band-saws and other large pieces of carpentry machinery driven by a gas motor, machinery for chemical factories, mining machinery, machinery for grain milling, locomotives and other railroad equipment, centrifugal pumps, kerosene motors, printing presses, steam-driven "locomobiles" (slow, tractor-like vehicles).[69]

BIRGITTA CONRADSON (1989), 12[70]

The advertisement ought, above all, be seen as a model of the current norms of a society. Advertisements, posters and advertising brochures are a part of our culture, but just as well a factor in the creation of [hegemonic] culture.

PHIL PATTON (1993), 43

Above all, the fair [Chicago's 1893 Columbian Exposition] was the dry run for the mass marketing, packaging and advertising of the 20th century.

1897 issue of *Aftonbladet*, a Stockholm daily newspaper, as quoted in Anders Ekström (1991b), 128

We live in the age of advertising and advertisements.

In the spectacular dreamworld of the Stockholm Exhibition the display of each exhibiting company was in itself an advertisement. But aside from the commodity placed on stage, and aside from the little building put up by Sofia Gamaelius' advertising bureau (Sweden's first), there were other floating images, marketing messages, that more or less constantly beset the visitor. At times one might be literally bombarded with advertisements, with advertising fliers released from a hot-air balloon that had lifted from in front of the Hall of Industry.

Almost wherever one went something free was to be had in the form of an advertisement, of an object projecting commercial meaning, of a souvenir meant to arouse nostalgia for the Exhibition and thereby a longing and desire for the goods encountered there.[71] At the Moorish Coffee House – where the building's architecture, its furniture, and the attire of the waitresses all linked the product with the exotic other, and thereby with foreign origins – the manufacturer of a surrogate coffee gave away the cup and saucer with which each customer was served. (Representation of the exotic other was very limited at the Stockholm Exhibition, unlike the situation at expositions held in the metropolises of the colonial powers, where entire "neighborhoods" or "villages" were represented. While in Stockholm the display of the exotic other was primarily an advertising ploy, it elsewhere served foremost as an expression of national power, as a declaration of moral and technological superiority, as proof of the genetic inferiority of non-European races, as another element in the exhibitionary rhetoric of progress, as a device to "mobilize support for the policies of national imperialism."[72]) At the Cloetta Brothers' Pavilion a free aluminum cup was distributed to each purchaser of hot chocolate. At another individual-firm pavilion a "cooling glass" of *Freja Nektar* was to be had without payment. At an ice-cream kiosk – perhaps the first in Sweden – a dessert glass came with each purchase. At display after display one might be given a small expensively printed catalog, a single information sheet, an engraved metal card as a reminder of specific metal products, a post-card, a take-home sample of some kind, a colored cardboard cutout congruent with the exhibitor's product or trademark. When purchasing a small object, one was apt to receive it in a bag of high-quality paper with the producing company's name in gold embossed letters. At the pavilions of the various breweries, as well as at some of the eating establishments, one was apt to be given a cardboard coaster or napkin advertising the beer one was consuming.[73]

In many, if not most instances, these various souvenir objects – as well as the advertising posters to be seen at some locations – linked the imagery of the company or product with the imagery of the exhibition. By juxtaposing a factory photograph or trademark with a picture of the Hall of Industry or a panoramic view of the exhibition grounds, these objects suggested that the products to which they referred were also memorable items, worthy of national pride. Nationalistic or patriotic sentiments were also repeatedly evoked by these objects through the use of blue and yellow colors; through depictions of the flag; through, where appropriate, making much of being either a "Purveyor to the Court" or the winner of one of the numerous gold or silver prize medals distributed by King Oscar II on September 23; or through finding some other excuse to employ a profile of His Majesty. (The sales to be derived from advertisements mentioning an exhibition medal were deemed so important that many exhibitors complained of jury incompetence or partiality prior to any decisions; while others, extremely dissatisfied with the bronze medal or "honorable mention" accorded them, placed newspaper announcements "To the Public" asserting the "Prize Judges" were themselves worthy of nothing more than "honorable mention."[74]) Not only was the King's profile exploited in candle-wax, china, marble and cast iron; but the Cloetta Brothers did not hesitate to draw attention to their chocolate products by anything less than a life-size chocolate sculpture of the King and Queen sitting on their thrones beneath an ornate canopy.[75] Commercial and nationalistic meanings were also repeatedly merged through the use of young women in folkdresses as display attendants, sales personnel or waitresses. If, as in most cases, the distinctive clothing worn by these women was from the province of Dalecarlia, it carried an "aura of reliability, quality and high status," an association with "strength, endurance, dutifulness," while at the same time "incarnating" the bourgeois national-romantic myth of the truly Swedish, of a "simple, secure and honest peasantry, living in a colorful, rich and unspoiled landscape"[76] of people living in a landscape devoid of class tensions.[77] (Because the image of the folkdress-clad Dalecarlian woman suggested both product quality and conditions of interclass unity, it also was frequently used in trademarks or posters for goods as different as chocolate and dairy separators.[78]) Moreover, as the young folkdress-attired women were often attractive, they were forerunners to a central cliché of twentieth-century advertising, their presence more or less subtly communicating a connection between product and sexiness or seductive capacity. In the Chemico-Technical Industries Building a woman in Dalecarlian dress openly flirted with men in an effort to pursuade them to purchase an "unrivalled anti-freckle pomade" for the women who were accompanying them.[79]

The international exhibitions of the late nineteenth century were meant to be glittering showcases. They were thereby like crystallized condensations of the major European capital cities of the period, cities that were meant to be like a glittering jewel in the national crown. They were thereby like miniatures of Benjamin's Paris. "Paris, a 'looking-glass city,' dazzled the crowd, but at the same time deceived it. The City of Light, it erased night's darkness.... The City of Mirrors – in which the crowd itself became a spectacle – it reflected the image of people as consumers rather than producers, keeping the class relations of production virtually invisible on the looking glass' other side.... [Paris] – a magic-lantern show of optical illusions, rapidly changing size and blending into one another."[80]

TONY BENNETT
(1988), 98

[T]he substitution of observation for participation was a possibility open to all [at the international exhibition]. The principle of spectacle – that, as Foucault summarizes it, of rendering a small number of objects accessible to the inspection of a multitude of men – did not fall into abeyance in the nineteenth century: it was surpassed through the development of technologies of vision which rendered the multitude accessible to its own inspection.

Yet the cultural conditions of seeing were starting to change, and the Eiffel Tower stood for that too. The most spectacular thing about it in the 1890s was not the view of the Tower from the ground. It was seeing the ground from the Tower.

ROBERT HUGHES
(1991), 14

If one speaks today with someone who experienced "1897," his eyes take on a glow. The memories well up . . .

MARTIN OHLSSON
(1959), 192

Back again into the world of dreams and illusions. Rise to the heights. Pay your fee and ascend one of the minarets of the Hall of Industry by way of an electrical elevator.[81] "... make a little air trip to the pinnacles of the temple.... be literally uplifted." Step off, look about, "involuntarily emit a cry of admiration before the glorious panorama." "An enchanting painting," "new beauties" whichever way one turns – the entire exhibition, beyond that the "dark house masses" and "glittering waters" of the multi-island city, beyond that the endless horizon.[82] Above all, see rather than be seen. Assume a detached, panoptic point of view. Know the knowledge and visual power of the powerful. Gaze at the cityscape beyond the exhibition; gaze at people

reduced to specks of insignificance, swallowed up in what illusorily appears as a harmonious and natural totality.[83] Intensely sense the promise of technology, "the promise of unlimited power over the world and its wealth."[84] Stare at what is immediately beneath, the halls and pavilions containing all the goods you can imagine. The entire world of consumption literally at your feet. A heaven on earth where everything is possible. An earthly paradise where there is no end to material happiness, to a now and future happiness made possible by national progress. Grown dizzy from the surrounding sights of bliss and harmony, grown giddy with visions of consuming desire, grown light-headed with fantastic images of the future, rush across a flying bridge to the central cupola in order to confirm that you have entered heaven, to inform others that you are sitting on top of the world, to write a "wish-you-were-here" postcard. Articulate modernity. (Buying and writing post-cards that could be mailed on the spot – postmarked *Kupolen* – became a "real mania." By the end of the exhibition over 100,000 cards had been purchased and sent from the cupola location, thus, in all probability, defining the initial popular breakthrough of the postcard in Sweden.[85] Less commonly, people confirmed their heavenly ascension by imprinting their name or initials on the safety railings.[86])

Observe from a less vertiginous height. Stand six to eight meters above the ground, on the bridgeway connecting the Hall of Industry and the terrace

of the Central Restaurant, or on one of two other such structures. Gaze at the people below, strolling from one building to another, riding carriages or open-air, horse-drawn trolleys. Gaze at their attire and parasols, at their social markers, at their bearing and gestures, at their actions. Be a spectator. Allow others to be part of the spectacle.

And now, from a little octagonal building at the foot of a tree-grown hillock just to the right of the exhibition's main entrance, or from an identical building behind the Machinery Hall, see and notice from ground level while being unseen and unnoticed. Eye the world from beyond the world's eye. From the darkness within the "camera obscura" watch the rotating angle-changing disc, watch its somewhat reduced but sharply defined reflections, watch it magically present a stream of images, watch fixed objects, human movement and facial features flow by in detail. Through the camera's all-seeing periscopal eye observe the intimate. Spy on others without their knowledge. Observe restaurant customers shamelessly flirting with waitresses. Observe the passionate kisses of a couple on a secluded bench.[87] Be titillated in these or other ways while simultaneously (re)learning the lessons of modern power, of panoptical bureaucratic power. Be reminded of the new technologies of controlling vision, of the ever-present eye of State Gods, of social guards. Be reminded that the power of the State lies not so much in its ability to rally police or locally stationed troops at points of actual or potential disturbance, as in the ability of its constituent agencies to keep an eye on, to maintain bureaucratic surveillance over, each and every subject. Be reminded that one may not always be out of sight just because it appears so. Be reminded of the bureaucratic mode of specular dominance to which one is constantly subject-ed. (At the time of the 1897 exhibition Stockholm was divided into over twenty *rotar*, or registration districts, each of them presided over by a bureaucratic administrator who kept his entire population under microscopic observation, noting not only in- and out-migration, births, deaths and marriages, but also school attendance, vaccinations, hospital stays, the joint residence of unmarried couples, changes in religious belief, arrests and court judgments, personal bankruptcies and other individual details.[88])

The subterranean environment is a technological one – but it is also a mental landscape, a social terrain, and an ideological map.

ROSALIND WILLIAMS (1990), 21

Participate further in the dialectics of the phantasmagorical dream. Move from the realm of above-ground sights and illusions to the realm of underground sights and illusions. Enter into another of those fantastic buildings, the pavilion of Scanian Large-Scale Industry, with its advance tower and differently colored glazed bricks that now and then produce a "strange, but certainly not unpleasant, glow." Watch through the glass windows of a descending elevator as the walls of a coal-mine shaft quickly flash by. Arrive in the underworld so as at one and the same time to seek the sublime experience and become a social voyeur. Recreate "one of the most enduring and powerful cultural traditions of humankind, a metaphorical journey of discovery through descent below the surface."[89] Renew the mythological quest for truths hidden in difficultly accessed subterranean regions. Venture for the truth about lower-class reality, about those who threaten with revolutionary upheaval, about those who dwell in poverty and thus evoke sympathy.[90] As you disembark from the elevator, be informed by a guide that you now are at about 100 meters depth. Wander about in an environment where the technological has replaced the natural, where "day has been abolished and the rhythm of nature broken,"[91] where consumption-enabling production is an around-the-clock business. Wander about in a labyrinth of low-ceiling passageways where "black diamonds" here and there shine in the dim light cast by a distant lantern and where extra props occasionally guard against the collapse of looser materials. Encounter a horse, under the control of a young boy, pulling a fully loaded cart along a narrow gauge track. Encounter headlamp-wearing miners wielding their pick-axes or using their own strength to move a cart. Encounter other miners and horses. All totally believable. All totally life-like. None of them living. All stuffed stuff – the living dead of the commodity-fetishism underworld. And the mine not a mine. Instead, an artificial creation but three meters beneath the surface. (The elevator had descended in ultra-slow motion, but a large rotating drum outside its windows had produced a whirl of images and thereby a falling sensation.)[92]

Follow the attracted flock. Stand transfixed in a lower-level room of the Stockholm Pavilion, before a full-scale cross-section model of the bowels of the earth beneath a central-city street intersection. Above eye level, a painting of buildings and street life. At eye level, the workings of the underworld laid bare, buried truths uncovered, the mysteries of the underground directly revealed and experienced: water pipes, gas pipes, cable drums, sewer pipes high enough for a fully grown adult to stand erect in.[93] Lesson learned: the technological progress of the nation is no mere surface matter, the products of industrial capitalism are fundamental to everyday urban life, are the means by which the modern city is regulated and unified, are the life-perpetuating blood vessels of modern existence. Contradiction silenced: the technological progress symbolized by the wonders of subterranean engineering requires that

countless men submit themselves to "the terrible realities of subterranean labor."[94]

Continue the dream tour of the magical, the magical tour of the exhibitionary dreamworld. Purchase your ticket and plunge into the murky *Sagogrottan* (Fairyland Cave), aboard a gondola-like boat, a vessel whose twin rear oars are powered by a standing young woman in Dalecarlian folk-dress. Glide from chamber to chamber beneath the mountain (a great wooden frame covered on the outside by tightly woven nature-like fabrics). Experience something "fabulous," a realm of contradictory illusions, a paradisiacal kingdom of darkness, a bewitching sequence of spaces where one automatically knows that the only proper voice is the whisper of awe. In one chamber mountains of silver rise out of the water. In another, walls are studded with gleaming gems, with diamond, ruby, emerald and sapphire replicas. In another, there are cascades of water in sparkling colors and lightning flashes in green, yellow, red and blue. In yet another, a "sea monster" makes a sudden appearance – the snake of this Eden of lighted darkness.[95] Thoughts on exit: "It glistened worse than a Thousand and One Nights – five minutes here was more beautiful than all those nights."[96] Thoughts inspired, sensations and illusion yielded, by the massive use of new consumer goods, by the use of

hundreds of colored electric-light bulbs of Swedish manufacture and huge quantities of tinfoil.

Outside, especially as the still-lengthy days grow perceptibly shorter in July, the evening-time dreamworld becomes an illuminated dreamworld, becomes images of light upon dark, of electrical enlightenment. The de Lavals Pavilion aglow with electric light, thereby publicizing its light-bulb factory. The St. Erik's Brewery Pavilion framed by electric bulbs. Tall electric lampposts flanking the approaches to the Hall of Industry as well as other walkways. Broad floods of light – "shining stronger than the sun" – cast by three rotating spotlights: one at the tip of the giant *Liljeholmen's* candle, the other two atop a tower of the Nordic Museum and a Hall of Industry minaret. In front of the Nordic Museum, a decorative electric-light message: "May Sweden flourish" (bloom, prosper, thrive). *La Fontaine Lumineuse*, or *Eldfontänen*, for twenty minutes "a spectacle as fabulous as the Fairyland Cave" – an enormous jet of water shoots up from the inlet in front of the exhibition, rising and falling to different heights, inspiring a collective gasp as it constantly

changes colors — now a cascade of silver, now flowing gold, now a liquid symphony in emerald green, now an explosion of red, now a shower of fantastic blue — by way of glass reflectors and an arc-lamp operated from within the cistern at its base. The "illumination evenings" that commenced in late August. At first only garlands of colored lanterns — 10,000 or more draped around the lawn edges and flowerbeds of the main esplanade — embellish the forms of electric light already present. However, a month later, during the Royal Jubilee week, the "illumination" extends to numerous pavilions and kiosks — with the King's initials and other anniversary symbols appearing on some of the larger buildings — and echoes with the magnificent fireworks lighting the sky from in front of the Royal Palace, a short distance away.[97] ... Prior to the exhibition electric lighting had been present in Stockholm; but never in such profusion, never in such brilliance, never in such concentrated form. (In 1897 Stockholm's trolley lines began to be electrified; but, still only one of about 7,088 public street lights was electric and a total of no more than 62,305 electric light bulbs were served by public and private utility companies.)[98] In the dreamworld of the exhibition electric light was the most pervasive testimony to Sweden's industrial and technological progress,[99] was the most insistent indicator of the possibilities of the dawning twentieth

century, was the quintessential expression of happiness ahead, of a bright future that all were to believe in. Here, in this magical light, was "the related [bourgeois] fantasy of a labor-free source of power," and therefore of a world free from social strife.[100] Here was Enlightenment, the application of scientific knowledge in the interest of human progress offered as popular religious spectacle in the interest of the powerful.

Did the opium fully work?

MICHEL FOUCAULT (1975), 25–6, as quoted in James Duncan (1990), 22

Memory is actually a very important factor in struggle ... if one controls people's memories, one controls their dynamism.... It is vital to have possession of this memory, to control it, administer it, tell it what it must contain.

MICHEL FOUCAULT (1972), 14, as embellished in M. Christine Boyer (1992), 191

[T]he muffled fiction of history is "a place of rest, certainty, reconciliation, a place of tranquilized sleep."

M. CHRISTINE BOYER (1992), 204

The private sphere of nostalgic desires and imagination is increasingly manipulated by stage sets and city tableaux set up to stimulate our acts of consumption, by the spectacle of history made false.

[In organizing the Paris Exhibition of 1889] the compelling appeal to [the] common historical legacy [of the French Revolution] was a means for transcending contemporary social division.	DEBORA SILVERMAN (1977), 73

What surprised the non-European visitors to the exhibitions was the realism of the artificial. The famous Rue du Caire at the 1889 Paris exhibition reproduced an entire street of the Egyptian capital [down to the detail of dirtying the paint on building walls], and imported real Egyptian donkeys and their drivers. By its realism, the artificial proclaims itself to be not real. The very scale and accuracy of the model assure the visitor that there must exist [or previously have existed] some original of which this is a mere copy.

The commercialism of the donkey rides, the bazaar stalls and the dancing girls was no different from the commercialism of the world outside.

TIMOTHY MITCHELL (1991), xiii, 10

Gamla Stockholm, Old Stockholm. By day or evening, the illusion of all exhibitionary illusions. The most popular of all illusions (12,000 or more paid admissions on high-traffic days). An illusion so popular that pressure mounted to retain it beyond the exhibition's end, to preserve its "shimmering dream images." An illusory space in which collective historical memory is to be shaped and buttressed through direct bodily presence, through the physical suggestion of authenticity. A dreamworld space, a time-capsule trip, in which a constructed national pride, a fabricated national heritage, an invented national community, is to be reaffirmed. A mythic space, set aside, to be approached by a wooden bridge. A walled city of the late sixteenth century, of the period around the end of the reign of Gustav Vasa – "the name and time which beams most beautifully with the clearest light in our history."[101] A city of towers, turrets and spires, of defensive cannons set within its outer walls. A city of winding streets and narrow alleys flanked by three- and four-storied renaissance buildings, many of them step-gabled, all of them painstakingly detailed. A city of classic landmarks – the Royal Palace and its Three-Crowns spire, the Town Hall and the big market-place in front of it. A city populated by people of the period: iron-helmeted soldiers; neatly dressed craftsmen, musicians, shop attendants and waitresses – nobody and no-thing ragged here to disturb the illusion. A romanticized imitation of the urban past imitating other romanticized imitations of the urban past. (Historical urban environments had been reconstructed earlier at international expositions in Antwerp, Edinburgh, Paris and, only in the previous year, Berlin.)[102] A city of maybe-I-am-not-dreaming illusions.... "The whole thing seems utterly amazingly

illusory;" "a stroke of magic;" "the highlight of the exhibition gave such an illusory impression ... that one was immediately swallowed up in the situation, a strange dream-being outside of oneself tramping forward on the historical ground."[103] "Everything appeared so old-fashioned, and was so true to life, that as one wandered about there one got the illusion that one truly lived in the sixteenth century – it was truly a beautiful and romantic illusion."[104] "Even the critical observer gets a vivid illusion of standing eye to eye with the past."[105] ... A city of solid foundations, cut granite and brick that was not a city of solid foundations, cut granite and brick, but a city of wooden structures covered by deftly painted woven fabrics and plaster of Paris, a city half of whose ground rested on posts sunk into inlet waters. A sizable and spacious city that was neither sizable nor spacious, but appeared so owing to the extensive use of clever perspective manipulations. An Old Stockholm that was not just an Old Stockholm; but a pastiche of building replicas – from other Swedish towns (Malmö, Lund, Örebro, Vadstena, Visby), from Sweden's former German possessions, as well as from the renaissance capital city itself. A city of the past that was a pastiche of the past, present and future, that was yet another exercise in the exhibitionary rhetoric of national and technological progress. ... Progress-ive glimpses of the now and the future. In a tower room, a popular lecture-demonstration of an X-ray device. In another building "living pictures" that "hopped and danced."[106] The *Cinématographe Lumière*, or *Kinematografen*, Sweden's first film theater, with its fleeting and fragmented qualities, a metaphor for modernity. A program that changed frequently, if not weekly. Twenty-minute programs that included fourteen or more discrete elements. In rapid succession a royal wedding in Denmark, Japanese jugglers, a swimming scene, the same scene shown backwards for laugh effects; or a variety-show number, followed by the demolishing of a wall, a scene of attacking troops in historical dress, and the Brooklyn Bridge; or the present in new light – the arrival of King Oscar II at the exhibition's opening ceremonies, jubilee highlights, and scenes from within *Gamla Stockholm* itself. ... A historical city before its time, a historical city as "postmodern" commercial theme park, a historical city as contrived site for paid entertainment and here-and-now consumption. ... Pay to see not only the moving wonders of the *Kinematograf*. Pay to join three or four hundred others at *Then Lustige Theatren*[107] (The Funny, or Comic, Theater) or the smaller *Bollhusteater* and there watch a performance of a play, quite likely in the national romantic genre, not unlikely dealing with kings and well known events of the Swedish past. Pay to enter *Marionett-teatern*, a miniature circus with clowns, acrobats, tight-rope walkers and performing dogs. Step inside buildings of the past to purchase products and services of the present and past: mass produced waffles (at "The Golden Waffle"), mass produced chocolates, candies and marmelades, cigars, gold jewelry, pewterware, glassware, antique objects and weavings, ceramic pots and vases, hair cuts, readings of the future. Pay to wine or dine Italian style at *Bellio's Taverna* or "The Artist's Shelter," Spanish style at "The Spanish Grape," Swedish style at "Saint Gertrud's Banquet Cottage" or in the vaulted cellar of the Town Hall, where the furniture and decor is "so true to the period that one finds it doubly difficult to reconcile oneself with the thought that this is a temporary creation doomed to disappear shortly."[108] Pay fifty *öre* to have your name or monogram quickly but artfully engraved on your watch or other personal belonging. Pay five crowns to acquire a finished portrait bust of yourself within twenty-five minutes. Pay for art that mimics modernity, that mimics the tempo of mechanized production. Pay at the beer-bar "Pleasant Home" for the fast food of the future, for the *frankfurterkorv,* for Sweden's first "hot dog" (sales figures repeatedly updated with pride in terms of kilometers sold).[109] ... Dreamily reeling in this onslaught of images and pinch-yourself-this-can't-be-real experiences, be further tempted to come again, to pay and pay again, by the hyper-festive atmosphere of the crowded streets, by the clarinet and bagpipe, violin and harp sounds of a trio of Italian street musicians, by the voice of an Italian troubadour, by the cheerful melodies and brass orchestral tones drifting out from Saint Gertrud's Banquet Cottage, by the quick-tongued jesters and agile-bodied acrobats performing here and there, by the staged mêlée outside the Town Hall among soldiers who have grown drunk with beer and raucous with refrains of the "Soldier-Boy's Song" ("we kissed the Pope's nuns and drank his wine").... *Gamla Stockholm* – history as a legitimating, deceptive transfiguration of the present. *Gamla Stockholm* – history as an illusory space idealizing the past so as to idealize the present. *Gamla Stockholm* – history as a wonderland space of the "good old days," as a comprehensible and uncomplicated space devoid of hunger and misery, of human subjects enmeshed in problematic social relations, of labor-versus-capital struggles. *Gamla Stockholm* – history as a hyperspace where a past, present and future collapse into one space, but where the everyday conflicts and cares of the present are all the same escaped. *Gamla Stockholm* – history as a fantasy space where the past is commodified, where what has been is reduced to a realm of purchasable objects and diversions, where the once was is packaged so as to provide trust and comfort to the present-ly disoriented, where every woman and man is welcomed into the generously padded bosom of the commodity form. *Gamla Stockholm* – history as a spectacular articulation of modernity.[110]

Late nineteenth-century international exhibitions were not merely object lessons in the commodity form, technological progress and national identity. They in themselves became commodified objects.

If the late nineteenth-century international exhibition was primarily a spectacle of the commodity form, it was also a spectacle that assumed the commodity form.

If the late nineteenth-century international exhibition involved a marketing of dreams, it was also a dreamworld that could be marketed, a dreamworld which – for a price – could be given objective, memorable substance.

If "the world exhibitions glorified the exchange value of commodities,"[111] they were themselves easily glorified.

Available at the exhibition, or elsewhere in Stockholm during and immediately after the exhibition, was the following assortment of items. The "original, elegant, cool" Exhibition Fan – "One side [a] map of the capital and the exhibition ... the other side designed for wall decoration."[112] A 355-page official exhibition catalog, a slimmer official guidebook and an unofficial guidebook. Hasselgren's (1897) massive "description in words and pictures." Two booklets of "Exhibition Songs;" sheet music for "The Exhibition Waltz," "The Exhibition March – Cheers for Stockholm" and "The Exhibition Festival March;" two piano albums and one song album of Swedish music – all having an exhibition theme on their covers. A comic-book-like illustrated narrative, *Trulls' and Anna Stina's Memories from the 1897 Exhibition. Pilbom at the Exhibition,* a small book of fictional memories. Postcards with panoramic views of the entire exhibition or with close-ups of specific buildings. Booklets with photographs of either *Gamla Stockholm* or an assortment of exhibition sights. Stereoscopic exhibition photos (sold together with stereoscopic viewers). A gold-colored relief of the Hall of Industry set within a cardboard frame. A decorative dish, in color, showing the bridge approach to the exhibition and the Hall of Industry. A coffee cup and saucer with exhibition motifs (identical with those which had been distributed freely at the Moorish Coffee House). An "Exhibition Spoon" in silver. Stamps already bearing the *Kupola* postmark (sold to philatelists). *Nordens Expositionstidning*, a newspaper solely devoted to the Stockholm Exhibition. Cigar cups, matchbox cases, watchcases, visiting cards and other objects in imitation gold or silver with one or another exhibition view engraved upon them. Similarly decorated ash trays, cigarette cases, cookie tins and egg cups. Medals with *Gamla Stockholm* on one side and a profile of King Oscar II on the other. And, numerous other souvenirs made saleable by the use of exhibition symbols.[113]

A number of these mass produced mementos, like many other items available at *Gamla Stockholm* and other exhibition sites, were truly kitsch, were objects of commercialized "pseudoart," were forms of momentarily fashionable art that – whether or not "tasteless" – asked nothing more of any working-class or bourgeois audience than its money. As examples of kitsch, as examples of what some have labeled as "the essence of modernity," as "cheap and not-so-cheap imitations" of what art was supposed to be, as "an aesthetic form of lying," as art imposed from above rather than art produced from below, these ideology-laden objects at one and the same time often provided their possessors with "an enjoyable illusion of prestige" and – through spurring the daydream, through reawakening memories of the spectacular – "a pleasurable escape from the drabness of modern quotidian life."[114]

From the song "For the 1897 Exhibition" as reproduced in Paridon von Horn and Fritz Gustaf Sundelöf (1975), 111

When strangers come here in masses,
there will be a flood of cash in our purses.

The Exhibition of 1897 was also a commodity in the sense that it was something to be "sold" to tourists, something to attract foreign visitors to Stockholm in particular and Sweden in general, something to draw Swedes from the rest of the country, something to entice outsiders to spend and consume in the hotels, restaurants and retail establishments of the capital city. Thus, in marketing the spectacle, the exhibition's "Press Bureau" did much more than merely disseminate information locally. A brochure, "Towards the North," was distributed to travel agencies in England and the United States. Another version of the same publication was spread to Berlin, Vienna and other major German and Austro-Hungarian urban centers. And, newspaper announcements were actively diffused throughout Sweden, while some 3,000 posters were freely distributed to general stores and other shops at about 800 locations outside Stockholm.[115]

SUSAN BUCK-MORSS (1989), 86, based on WALTER BENJAMIN (1972–89), V, 250, 253, 267

[At world's fairs and international exhibitions] the commerce in commodities was not more significant than their phantasmagoric function as "folk festivals" of capitalism whereby mass entertainment itself became big business.... In 1867, the [Paris] fair's fifteen million visitors[116] included four hundred thousand French workers to whom free tickets had been distributed, while foreign workers were housed at French government expense. Proletarians were encouraged by the authorities to make the "pilgrimage" to the shrines of industry, to view on display the wonders that their own class had produced but could not afford to own, or to marvel at machines that would displace them.

I was only a poor youth then, and a common laborer, and I was doing military service that summer, so my finances did not allow me to visit the exhibition more than a couple of times.

AXEL ADOLF OLSSON (1947), 1

By the time the Stockholm Exhibition had drawn to a close, it had attracted at least 1.5 million visitors. While this total was impressive in some respects – being more than five times the then population of the Swedish capital – it was dwarfed not only by the 32.4 million who went to the 1889 Paris *Exposition Universelle*, the 27.5 million who ventured to Chicago in 1893 to see the World's Columbian Exposition, and the 48.1 million who were to go to Paris in 1900; but also by the 6.0 million who visited the Brussels *Exposition Internationale* in the very same summer of 1897.[117] Because of the greater dimension of international expositions held elsewhere, and because of Stockholm's peripheral location within Europe, foreign visitors to the Stockholm Exhibition of 1897 only accounted for a small fraction of total attendance – despite the siren calls issued by the "Press Bureau." However, the domestic subtotal, equivalent to about 30 per cent of Sweden's population, served as something of a mirror of Stockholm's intensifying railroad- and steamboat-based economic articulations with the remainder of the country,[118] as an indicator that Stockholm was now truly a capital city in both senses of the term.

Satirical causerie in a pro-labor newspaper:[119]

> Per: *The Nya Dagligt Allehanda* [a conservative Stockholm daily newspaper] claims that the ongoing art and industrial exhibition is not an ordinary exhibition.
> Pål: What is it then?
> Per: It's a jubilee exhibition especially arranged for the glorification of the anniversary of our King's accession to the throne, claims our royalty-fawning contemporary.
> Pål: Oh! But parliament and the city government have of course appropriated significant sums for the exhibition. Have these appropriations only been made in order to glorify our Majesty the King's jubilee?
> Per: Yes, perhaps so.
> Pål: But what say the tax bearers, who must pay the music but don't even have the means to take part in the whole thing?
> Per: Who do you think cares about them? *"Leben and leben lassen,"* that's the day's password, you see.

Nordens Expositionstidning, June 4, 1897

The exhibition's administration is overrun with petitions to reduce the price of both one-day tickets and full-season passes.

If the consumption and other object lessons of the exhibition really were to be made available to those beneath the bourgeoisie, to the urban working classes and the country's sizable rural proletariat, then something had to be done in order to place the costs of attendance within their reach, in order to make their dreamworld education an economic possibility. The working-class population of Stockholm could be attracted relatively easily, for they could attend without incurring either long-distance travel or lodging expenses. Thus, as local popular demand grew, a series of price-reducing measures were taken. Commencing with the third Sunday of the exhibition, "People's Sunday" prices were introduced, the cost of regular admission being cut from one crown to 50 öre. In order to encourage some weekday attendance on the part of wage laborers, ticket prices were cut to 50 öre after 6 p.m. from August 23 onward, and after 5 p.m. for the exhibition's final nine days. Ten Sundays and holidays were also designated as "joint-traffic days," days upon which one crown purchased a general admission ticket that also could be used to gain entrance to the adjacent *Skansen* outdoor museum, to *Gamla Stockholm*, the Tourist and Sport Pavilion, theater performances and other exhibition features that otherwise charged a separate fee. Moreover, "family discount tickets," granting six admissions for three crowns on any day of the week, were eventually made available to male heads of household upon submission of a written request. And, in mid-July, the price of a full-season pass was reduced 50 per cent, to ten crowns; a price which still must have appeared exorbitant to most Stockholm laborers.[120]

Subsidizing measures of another magnitude were necessary if any but economically comfortable Swedes were to journey to the exhibition from beyond Stockholm; if any but members of the non-local bourgeoisie were to be taught to conflate nationalism, consumption and images of a bright future; if non-local industrial and agricultural wage-earners from even "the most distant settled areas"[121] of the country were to be made subject to hegemonic representations, were to be supplied with the makings of a national collective memory. Through combining monies provided by the State with local donations, each of the provincial and large-city sub-committees of the exhibition bureaucracy were able to award "travelling fellowships" to a number of "good and experienced" laborers (5,489 such fellowships were granted in the country as a whole). The State portion of the subvention varied from five to 30 crowns per worker; and, in a typical case, each worker-fellow received a free roundtrip rail ticket, 14 crowns in cash, a coupon worth 6.53

crowns to be cashed in Stockholm, and six exhibition admission tickets.[122]

One month after opening day, the exhibition's administration committee set about arranging so-called "People's Trains," or special-departure trains consisting entirely of third-class and so-called "troop cars" (specially converted freight cars). These trains offered exceptionally low fares, which included a general admission ticket, tickets to some separate-fee exhibition features, a meal at the "People's Kitchen" on the exhibition grounds and, in some instances, one night's lodging.[123] The trains, which eventually brought 11,000 visitors, were explicitly intended to enable "Swedish men and women of the so-called working classes to visit the exhibition for a relatively low price and to take advantage of what for them, perhaps more so than for many others, combines business with pleasure."[124] Being swept up with the fever of the exhibition, even the editors of *Social Demokraten* declared that the People's Trains were "a little effort in the direction that just we have always consistently called for: the fruits of civilization for *everybody*."[125]

By far the largest number of subsidized working-class visits to the Stockholm Exhibition took the form of employer-paid trips; trips that were encouraged by an administrative decision to sell factory and workshop owners blocks of 100 or more tickets at a cost of 25 öre each. The 70,000 or more local and non-local laborers who were able to use these block tickets at no expense to themselves received a variety of benefits in connection with their visit.[126] Most employers intent upon exposing their laborers to the image-messages of the exhibition dreamworld did not simply pay for rail or steamboat transportation to Stockholm where necessary, but in addition provided one or more of the following: the travel and admission costs of accompanying family members; lodging for one night; full pay for missed days of work (the block tickets were not valid on Sundays); spending money ranging from two to 20 crowns; tickets to *Gamla Stockholm* and other separate-fee events; free meals at the "People's Kitchen" or, more rarely, at one of the exhibition's better restaurants; or a two-crown jubilee coin which quickly attained a much higher value because of the limited number released. Estate owners and large-scale farmers (who were not eligible for the purchase of large blocks of discount tickets) in a few instances also provided similar "extras" to those of their agricultural workers who were given a free trip to the exhibition.[127]

[The Great Exhibition of 1851] transformed the many-headed mob into an ordered crowd, a part of the spectacle and a sight of pleasue in itself.

TONY BENNETT
(1988), 85

> *The order during the People's Train trip as well as in Stockholm was exemplary.*
>
> <small>An observer from a provincial newspaper as quoted in *Nordens Expositionstidning*, July 16, 1897</small>

For many of the bourgeoisie, both in Stockholm and Sweden as a whole, there was not a little anxiety about intermingling with the working classes on the streets or in other everyday public settings.[128] What the bourgeoisie termed the "under class" was seen as threatening not only because of their economic and political demands. They were, in addition, often regarded as virtually "another race," a less developed, "crude," "coarse," unclean and dangerous people, "a seething mass, a formless ... rabble," with many features that seemed to unite "them with the primitive, the animal in the world" – "insufficiently controlled impulses, less elevated wants, simpler pleasures, and so on."[129] If these elements of the population were to be brought to the exhibition in large numbers for educational purposes, then they preferably were to be brought under disciplined conditions, under conditions in keeping with the bourgeois precept of public order and (self-)control at all times. Thus, those travelling by the People's Trains were accompanied by one or more "monitors" – sometimes members of the police – who designated a "deputy" in each car to assist in the maintenance of order. Upon arrival at Stockholm's Central Station both "travelling fellows" and People's Train passengers were greeted by "monitors" assigned by the exhibition's administrative apparatus to watch over and guide them during their stay. Those entering the exhibition by way of a block ticket had to be joined by a foreman or someone of higher status within the sponsoring firm. (Some companies made it easier to keep an eye on their employees by insisting that they wear a bow-tie or a chest ribbon decorated with the firm trademark, thereby also converting the men and women in question into walking advertisements.) At the end of the day, after being enlightened by *La Fontaine Lumineuse*, People's Trains visitors who had travelled all of the previous night were assembled and marched off in military fashion back to the Central Station where another night of uncomfortable train travel awaited them.[130] Meanwhile, "travelling fellows," company-sponsored visitors and end-of-the-season People's Train passengers who were remaining in Stockholm overnight, were put up at one of two barracks where they were subject to military discipline and held to a long list of rules regarding cleanliness and order.[131]

For those who all the same wished to minimize contact with the "under class," who did not even wish to encounter "monitored" laborers while at the exhibition, there were a number of strategies of temporal and spatial avoidance that could be employed. One could visit between 8 a.m. and 10 a.m., when admission was limited to those who either had a full-season pass, or were willing to pay two crowns rather than the customary one crown or 50 öre. (The latter option apparently did not prove popular except on September 23, when there were special ceremonies in connection with the Royal Jubilee. Except for that date, only 942 individuals took advantage of the two-crown alternative during the entire season. However, as there were well over 5,000 full-price season-pass holders in Stockholm, the number of early morning visits may have been considerably larger.) One could explore the exhibition late in the morning or during the afternoon on a weekday, when time–space segregation of the classes might not be total owing to the presence of "travelling fellows" and block-ticket vistors, but when one could avoid the largely unsupervised masses – of sometimes 30,000 or more – that appeared on Sundays. If one chose, one could make one's way to the *Djurgården* site of the exhibition by way of a horse-drawn cab and thereby avoid mixing with common folk *en route*. (Three new strategically located cab stations were opened in May, 1897, for the explicit purpose of easing access to the exhibition.) And, when it was time to eat, one could enter one of the fancier restaurants on the premises, where price levels precluded the presence of working-class women and men.[132]

Evenually, however, the fear of class intermingling appears to have diminished. Quite soon, in fact, it came to be recognized that virtually all of the "under class" women and men who showed up at the exhibition did so in their Sunday best, rather than in their often shabby everyday work-clothes; in their beflowered hats, ground-length skirts and dresses of white, black or pastel color; in their dark suits, white collars, vests, cravats and black or straw hats; in their I'm-just-as-good-as-you attire (an attire that did not hide the origins of those whose bodily bearing was less upright, less uptight, than that of the bourgeoisie). Eventually, it could be comfortably observed that: "One has the drama of a many-colored throng before one's eyes, and that is no means the worst spectacle one can be offered."[133] Eventually, it even became something of a "sport" for bourgeois women and men to take an afternoon meal at the "People's Kitchen," an institution set up to accommodate people of "lesser means" with "truly cheap" meals – generously portioned meals that cost either 40 öre (soup plus meat dish), or 30 öre (meat dish only). Or, at least it became common on weekdays; for, on Sundays, when People's Train passengers had to be fed, and as many as 3,500 meals had to be dished out, the realities of a wait of several hours among a sweating and impatient crowd could not have proved especially appealing to bourgeois sensibilities. Modelled after a "steam kitchen" that supplied food to Stockholm's poorest during the winter, the People's Kitchen provided meals that were prepared with factory-like efficiency according to principles of "scientific rationality." These meals were served by a crew of "society-lady" philanthropic volunteers in a dining-room whose walls were in keeping with the commodity-form's saturation of the exhibition, as they bore the large-signed advertisements of food suppliers. It was apparently the fact that King Oscar II – the exhibition's living icon – had himself taken a meal at the People's Kitchen which convinced many that class mixing (and money saving) at this particular site was not an action beneath their dignity, even if the women serving them saw themselves as charitably missionizing the working class.[134]

Like the consumption goods and commodity forms upon which they centered, the international exhibitions of the late nineteenth century were ephemeral. Being quintessential expressions of modernity, being spectacular articulations of modernity, they were always temporally and spatially short-lived. With a duration that rarely extended beyond the warmer months, they were like an intense version of any fleeting summer fashion, any short-lived fad. With most, if not all, of their buildings designed for quick dismantling or demolition, their landscapes were like prolonged summer reveries, their geographies here today and gone after not too many tomorrows.

When all is considered, the secret of capitalism's revolutionary capacity, the secret of capitalism's repeated self-reinvention, the secret of capitalism's ability to survive crises of over-investment and excess production capacity, lies in acts of creative destruction.

When all is considered, claims Rydell, what the international fairs of the half-century preceding World War I offered "to millions of fairgoers in the wake of … industrial depressions and outbursts of class warfare" was a "cohesive explanatory blueprint of social experience."[135] They were offered what Berger and Luckmann (1967) term "'a symbolic universe,' … a structure of legitimation that provides meaning for social experience, placing 'all collective events in a cohesive unity that includes past, present and future.'"[136]

When all is considered, it is the articulations of the taboo-breakers that most clearly reveal what is most central to their time and place.

Columnist in *Dagens Nyheter*, October 5, 1897

Is it all over already?

Consider the spectacular end, the dreamworld climax, the October 3 closing of the Stockholm Exhibition, the final pronouncements on the pronouncedly different patterns of consumption that had been represented there. Unlike the opening ceremonies, this is not an event restricted to the socially and politically well placed, to the economically well heeled. Here no royal splendor and gaudy diplomatic corps, but people of varying backgrounds in their best attire. Here no ten-crown admission barrier, but 50 öre tickets. Here no class segregation and limited numbers, but an open celebration with an estimated 60,000 or more in attendance. Arriving by horse-drawn cab or steam-launch, by trolley and by foot, women and men from the working classes as well as the low and high bourgeoisie immerse themselves in illusion and phantasmagoria one last time. Every pavilion and restaurant is filled to capacity. The queue to *Gamla Stockholm* appears endless. The throng purposefully in constant motion to see what was missed on earlier visits, to see what must be seen once more. At 5 p.m., the closing hour for all buildings, a warning gong sounds from within the Hall of Industry, the swarming crowd collects in front of that building, the Svea Guards military band strikes up "Hear Us, Svea," a "touching" national-romantic hymn. A final brief speech delivered by Exhibition General Commissioner-Governor-Baron Tamm, with other Central Committee members and exhibition functionaries standing behind him. Key articulations: "Lastly, the exhibition administration wishes to convey its thanks to the public who visited the exhibition and who so strongly contributed to its turning out with good economic results. It may be counted as a great honor to this crowd that no troublesome disturbances have occurred. The seriousness which has rested over the exhibition has certainly not excluded happiness, but happiness has never degenerated into frenzy and rash behavior. For this – much obliged! ... this exhibition which for many people will only be a memory, but for yet more others has provided learning. . . . Herewith the exhibition of 1897 is concluded! Let us all unite in the customary wish: God save the King and the Fatherland!"[137]

All hats off now, blaring fanfares from the minarets, the boom-boom-boom of saluting cannons, cheers upon cheers, cheers from below echoed by cheers from the heavenly bridges above, the military band resumes, a boisterous singing of the national anthem, a slow lowering of the flag high above the main cupola, yet more cheers upon yet more cheers, from somewhere within the crowd – "A four-fold cheer for the Governor" – the sounds of a march.

AND THEN, immediately afterwards, a final spectacular articulation. Bourgeois taboo broken. Everything no longer in its time and place. Rampant plundering. Unchecked looting. The spontaneous appropriation of objects, of commodities, of "souvenirs." The "Day of the Kleptomaniacs" (title of the cartoon reproduced here from *Söndags-Nisse*, October 10, 1897). Everybody into the act – "elegantly dressed women" and "finely attired businessmen" as well as drunken youths who occasionally used their canes to smash what they couldn't take. All under the eyes of exhibition watchmen who offered no protest, although the municipal police eventually made a few arrests. The peek-a-boo end of illumination: first, 10,000 or more colored lanterns and

bulbs before your eyes; then, Allllllll gone. Chinaware and glassware hungrily grabbed. From one establishment alone, 7,000 dishes and 2,000 glasses surrendered. From another two restaurants about 409 dozen glasses, 200 large beer mugs and 180 dozen dishes relinquished. Other personal "annexations": coffee pots; silverware; sugar and cream sets; at least one cannonball from *Gamla Stockholm*; chairs and a variety of carryable items that had been a part of individual displays. "Everybody had something to carry – enormous candles, sign plates, ceramics and bottles were most numerously represented – but the proudest to be seen were those who bore placards attached to their highly held canes, printed with the words 'Gold Medal.'"[138]

Proud editorial claim one day prior to the drama of October 3: "[The exemplary behavior of visitors throughout the exhibition] is quite simply admirable and ... [validates] the King's recently made observation that the Swedish people are the most mature people of our time."[139]

Joke one week later:

A. Do you think it will be long before they take away everything from the exhibition?
B. Oh no, it's going quickly, but if they arranged an extra closing day it would be done in a single evening.[140]

And what was the meaning of all this, of this final-day "orgy," of this "excessive" "mob" outburst?

What meanings were here being enacted or constructed by women and men of heterogeneous social origins and class circumstances, by women and men of varied geographical background, by women and men of dissimilar age and biographical experience, by women and men whose predispositions and identities were far from uniform, by women and men otherwise inclined to read situations very differently?

What, if any, were the unifying meanings?

What were the vocabularies that gained spectacular articulation here?

Was this the final step of the pilgrimage? Was this a collective act of communion? Was this a confirmation of complete consumer conversion, of unreserved belief in the sign of the good, of total surrender to the new consumption religion? Were the objects removed, the objects swallowed by the hand, the flesh of the fetish Commodity, the blood of the Almighty God Commodity? Was petty theft thereby the production of pure meaning?

Had the educational purposes of the exhibition been fully fulfilled? Was this a collective declaration of lessons learned, of lessons learned all too well? Had the gazing pupils moved from the passive absorption of theory to the over-exuberant, all too perfect, exercise of practice? Had the students of the new technologies of vision seen the light, read the exhibition's landscape of desire all too clearly? Had the new dominant ideology of consumption been consented to all too enthusiastically by subaltern students? Had messages and images that glorified industrial accomplishments, that glorified Swedish goods, been absorbed without question, been absorbed to the point of saturation? Was the appropriation of property an appropriation of the meanings offered by the exhibitors? Had the boundless profit desire of exhibitors been (over)successfully translated into a boundless, uninhibited desire to possess?

For how many was this not an acceptance of meanings provided by others, but a manifestation of discontent, a way of struggling over those very meanings? For how many was it a matter of rejecting exhibitionary schooling, of having had enough of "Look, but don't touch"? How many lived in circumstances that left much or all of the newly offered pattern of consumption beyond their reach? How many lived in circumstances of everyday need, power-relation subordination and humiliation that led them to disbelieve the pervasive representations of progress and national pride, to be incredulous of the imagery of an ever-brighter future? For how many was it a question of refusing to wait, of grabbing something right here and now – before it all disappeared?

For how many was it a matter of simultaneously rejecting the Oscarian/Victorian lessons offered in the unspectacular school of everyday life; of refusing to become self-controlled in the face of temptation; of refusing to become a self-regulating norm-alized citizen, always able to distinguish between correct and incorrect, proper and improper; of refusing to exercise self-discipline in all situations; of refusing to avoid emotional outbursts and dis-order at all times and places; of refusing to submit to a would-be hegemonic discourse?

For how few of rural origins or current rural residence was this a confused response to a spectacular shock of modernity, to a set of concentrated meanings that was alien to personally central meanings acquired in the past? For how few was this an act of disorientation, an unreflected response to an inability to deal with the puzzling newness of the new, an unreflected response to a profound sense of disjuncture, to the sensing of an unfathomable gap between an earlier or current life of rural need or pauperdom and the dreamworld of commodity-form meanings? For how few was this the spontaneous crystallization of an identity crisis; a matter of: What am I, doing here? Who am I, doing here? What is the (new) difference? Whether I do? Or I don't?

For how many was this a response to provocation? How many just could not swallow a compliment from Governor Tamm, especially a compliment for having maintained bourgeois orderliness, a compliment for not producing "troublesome disturbances," a compliment for avoiding "frenzy and rash behavior?" How many gagged on the words of the Governor of Stockholm, a man personally appointed by King Oscar II, a man who shared the King's aversion to the labor movement, a man who enabled the King to intervene in local politics? How many would just not take this from an inflated puppet of the King, a King whose portrait one would just as soon use as ass-wipe?[141] For how many was it a question of: "*Fy fan*!, [God damn it!] we'll show you what 'frenzy and rash behavior' are all about! We'll make object lessons of our own! We'll take these enlightening objects for ourselves!"

For how many was this a final engulfment in the dreamworld, a wish-fulfilment dream lived out, an uncensored action before waking up, a latching on to the phantasmagorical before it evaporated, a seizure of the illusion, a fantastic transgression to be acted out and then repressed in the collective unconsciousness? For how many was this the final act of the dream play, a case of being swept up in the vertigo of exhibitionary "reality," of taking action while experiencing the maelstrom of modern existence in extreme form, of acting while standing "at the precipice of the modern whirlpool where identity and reality are constantly created and dissolved anew."[142]

For how many was this final immersion in the dreamworld a death-denying dream, an anxiety-avoiding dream, a refusal of the abrupt ending, an attempt to capture physically a wonderland time before it ran out, a desperate gesture in keeping with the logic of modernity – that "time [which] doesn't want to know death."[143]

For how many was the dream content of this joyous pilfering of yet another type? For how many had the bourgeois dream of democracy been censored out a fear of the working class in power, been unconsciously reworked into a dream of total freedom to consume?[144] For how many did the looted object serve as a bridge between the conscious and the unconscious? For how many was the stealing of a kitsch-object one last kitschy escape from everyday realities?

For how few was this a loss of consumer innocence, a loss of innocence inspired by the exhibitionary dreamworld's kaleidoscope of seductive images, a loss of innocence unconsciously precipitated by the landscape's profusion of phallic towers (powers), a number of them – such as those of the Forest (Industry) and *Finspång* pavilions – shameissly obscene, unambiguously suggestive, in their heady anatomical accuracy? For how few had the accumulation of desire during repeated dreamworld visits reached a level where some final outlet was necessary; a level of overstimulation demanding release; a level where an all-consuming desire had to get lost in an object, any object that offered no resistance, that offered instant gratification?[145] For how few was this nothing more than an innocent submission to the irresistible, an innocent ravishing, nothing more than the snatching of "only a souvenir?" (words from a song, "Promemoria," which appeared shortly after the end of the exhibition and which were to be sung to the melody of "I Remember the Sweet Times"[146]).

Was this nothing more, for some, than an act of creative destruction, an imaginative act of pilfering and vandalism, to be carried out anonymously, in the safety of numbers, knowing full well that the exhibition's halls and pavilions were themselves to be creatively destroyed almost immediately? Were there not those who drew a parallel between the fleeting existence of the soon-to-be-wrecked buildings around them and the fleeting existence of the commodity? "In these terrible housing-shortage days, when so many families are without a roof over their heads,"[147] were there not those who were reduced to rabble by the knowledge that the Hall of Industry and other massive structures were to be torn down, reduced to rubble, and put to the uses of capital? (The contract awarded to the construction firm that built the Hall of Industry stipulated that all of the material used in this – the world's largest wooden building – would revert to the firm after October 3rd. Elsewhere on *Djurgården* the firm was allowed to establish a saw-mill and box factory so as to convert the Hall's beams and boards into packing cases.[148]

For how few – about to have their dreamworld stripped away – was this a reappropriation of space; a reclaiming of space that had been "ours" to wander about in, to play and picnic in, to take pleasure in, on pre-exhibition Sundays? For how few was this meant to serve as a reminder to those above that public spaces are not reserved solely for them, that public spaces are perfect sites upon which to contest cultural hegemony with new tricks and tactics?[149]

When all is considered, did the creator of the here repeatedly shown looting cartoon come closest to the mark, to fundamental meaning, when he depicted people as stolen goods; when he included Carl Bendix (head of the exhibition's everyday administration), Thore Blanche (the exhibition's press chief) and the Italian troubadour from *Gamla Stockholm* among the chairs, tables, candles, light bulbs and other pilfered items; when he "thingified" the subject; when he thereby – wittingly or unwittingly – equated the episode of riotous "kleptomania" with the triumph of commodity fetishism?

Was the closing episode of unbridled looting an over-determined metaphor for the entire Stockholm Exposition of 1897?

The *ex post facto* legitimation of a hegemonically controlled spectacle always requires wishful fabrication, if not outright calculated lying.

The earlier absence of unruly behavior at the exhibition claimed by Governor Tamm and the editor of *Nordens Expositionstidning* was something of a myth, an articulation with blinkers, a rhetorical ploy in keeping with would-be hegemonic discourse, in keeping with the ideal-ogic of absolute self-control and public-space order – especially at a spectacle of such national importance. A partial inventory of exhibitionary transgressions large and small: the firing of several missed revolver shots in the midst of a dispute at the Machinery Hall; the theft of a silver tray, "a real work of art ... valued at 300 crowns;" the theft of the King's binoculars; the theft of three revolvers, a gold watch, a gold medal, furs, weavings, expensive photography frames, engraved binders, ebony canes, brooches, jeweled pins, boxes of chocolate, books, penknives, cigar mouthpieces; the theft by "a wealthy young woman" of clothing pattern books and other small items; the "regrettably frequent" defacing and vandalizing of displayed consumer goods and machinery; countless acts of pickpocketing by "a number of individuals" – Swedes as well as "foreign adventurers;" the abuse of group and seasonal tickets; the constant occurrence of people "forgetting" to leave the grounds at night; the consumption, by a night watchman, of five liters of aquavit displayed in a glass anchor; a number of individual and collective drunken disturbances of the peace; the punching of two uniformed guards; the possible murder of a man who got separated from his acquaintances in a crowd and eventually turned up floating in the water.[150]

Every way of seeing is also a way of not seeing.   HELEN MERRELL LYND
as quoted in Stuart Ewen (1988), 156

Struggles over what will count as rational accounts of the world are struggles over how to see.   DONNA HARAWAY (1991), 194

By controlling the public stage, the dominant can create an appearance that approximates what, ideally, they would want subordinates to see.   JAMES C. SCOTT (1990), 50

Silence itself ... is less the absolute limit of discourse, the other side from which it is separated by a strict boundary, than an element that functions alongside the things said, with them and in relation to them within overall strategies.... There is not one but many silences, and they are an integral part of the strategies that underlie and permeate discourses.   MICHEL FOUCAULT (1978), 27

It is true that just as taboo exists in order to be broken, so transgression fortifies the taboo.   MICHAEL TAUSSIG (1993), 126

———

Where taboos are publicly broken by large numbers of people normally not thought of as taboo-breakers, those with the power of the word (who support those in power), those with the power to disseminate knowledge, may choose to defuse the situation, to discourage reoccurrence, by various strategies that wed silence and denial, rather than by strategies of condemnation or vilification.

To be publicly silent on what is known by word of mouth is to deny that transgression really has occurred, is to attempt to deflate it, is to attempt to diminish its significance, is to attempt to deny that it in any way threatens those in power. Such attempts, such forms of censorship, however, merely emphasize how central the taboo in question is to the idea-logics of those in power.

To deny publicly that a taboo has been broken, to give events another twist, to attempt to make something else out of them, to attempt to rationalize them, is to attempt to silence any thought that the transgression in question is ever really permissible, is to attempt to silence any thought that transgression really may be tolerated, is to practice hegemonic repress-entation.

———

Many public commentators and newspaper editors expressed disapproval, disappointment and disgust with the mass looting that marked the exhibition's closing. Such condemnation, however, was far from universal.

Like every other Stockholm newspaper, the liberal *Svenska Dagbladet* gave a florid and rather sentimental account of the exhibition's final-day atmosphere. But of the thousands of acts of petty pilfering there was not a single word. Instead, one day later – and perhaps horrified at the thought that some of its own readership was involved – an agile act of Freudian displacement, of concealing censorship, in the form of a story that began: "They have been taken from every fringe. Both precinct police and detectives have been out and about in order to pull in [street arrest] the many male and female [petty criminal] individuals who may be reckoned as dangerous to public safety."[151] The story was coupled with two other reports: one of a watch theft in broad daylight, the other of a boy who cut himself after stealing two bottles of beer from a delivery wagon. Despite the silence, the message was clear.

The printed voice of the labor movement, *Social Demokraten*, also waxed florid and sentimental over the final day's "jovial and pleasant" atmosphere,[152] also chose silence regarding the plundering rampage. Here too, an agile act of editorial displacement, most likely produced by thoughts of readership participation, by fears of casting members of the working class in a bad light, by the incompatibility of things as they happened and the unmistakably bourgeois values of the Social Democratic leadership. Here, after two more days, was a story focusing on the plight of the exhibition's uniformed watchmen, who elsewhere had been charged with standing by passively or even offering encouragement as the stealing riot unfolded. Here no mention of those charges, but an account of their discontent, of their having been underpaid, of their having been forced to purchase a trophy for their commander on his birthday.[153] Here the event text of October 3rd erased, here a subtext of labor-versus-capital struggle in its stead. (Earlier in the summer, *Social Demokraten* had given voice to the watchmen's complaints of long and fragmented working hours without opportunity for rest.[154] Elsewhere the trophy given the commander was described as "large, [and] singularly magnificent," inscribed with the words "From thankful watchmen."[155])

Hasselgren, the chief chronicler of the exhibition, like a few others, chose to deny by way of invoking "extenuating circumstances." On the one hand, the "excesses" committed "mostly by people of whom one expected more self-control and more respect for property rights" were "a dark spot on an otherwise bright picture."[156] On the other hand, this regrettable aberration had precedents. Twenty years before, when a restaurant closed in the southern part of Stockholm, its owner encouraged his guests "to take as much as possible home with them so as to 'facilitate the demolition of the building that was to begin the next day.'"[157] And, in connection with "several" subsequent restaurant closings, those present thereby saw themselves as having the right to "annex" "mementos." In short, the evening of October 3rd was more a matter of "tradition," more a "bad habit" to be prevented in the future than a true taboo transgression worthy of "a lot of noise."[158] That this was a bit far fetched, that no more than a very few – if any – of the plunderers could have been at *Björngårdskällaren* two decades earlier, or at one of a small number of restaurants on subsequent dates, was no obstacle to the rationalizing rhetoric of denial.

The "Official Report" of the exhibition, a two-volume 1,072 page account authored and issued by those who had organized the spectacle, opted for a form of underlined silence. On the very last page, where the last hours of the exhibition are described, only the words of praise and thanks regarding the absence of "troublesome disturbances" are quoted from Governor's Tamm's closing-ceremony address.[159] And beyond that, nothing but "God Save the King and the Fatherland" and total silence. Total silence as the final authoritative word. Total silence as a final spectacular articulation of modernity.

Afterwards, after the closing of the international exhibitions, after their dream-worlds have been physically dismantled and removed, after the consumption forms they dreamily mass marketed are no longer fashionable, after other new goods and fashions have come and gone, after the commodities they promoted have become but discarded props, those discarded props provide "material evidence that the phantasmagoria of progress had been a staged spectacle and not reality."[160] Instead of progress, the modern condition, one dream followed by another, and "the [dream] fashions of the most recent past [experienced] as the most thorough anti-aphrodisiac that can be imagined."[161] And for some – especially among those for whom the dream has repeatedly proved inaccessible – disillusionment, recognition of the falsity of the dream.

Afterwards, a final echo of the 1897 exhibition, an articulation of another kind. Stockholm, 1901. The bright future of the twentieth century no longer on the horizon, but present and grey, right here and now. Economic downturn. Falling wages. Hard times for many. And yet, new patterns of consumption constantly promised, new dreams constantly offered.

Opening words from the lyrics of "Falseness" (*Falskhet*), a song-sign of the times:

Wherever one looks – everywhere one can see,
false goods for which one gives good money.[162]

FIGURE ACKNOWLEDGMENTS

All photographs courtesy of *Stockholms Stadsmuseum* (The Stockholm City Museum).

NOTES

1. For further elaboration of this line of reasoning see Pred and Watts (1992). A somewhat different bridging of the two senses of articulation is suggested by Stuart Hall (Grossberg, 1986; Hunter, 1988, pp. 116–99).
2. Gustafson (1976); William-Olsson (1937).
3. Hammarström (1970).
4. Pred (1990a).
5. The "always-again-the-same" quality of consumption under modernity is one of the principal themes of Benjamin's *Passagen-Werk* (Buck-Morss, 1989).
6. Myrdal (1933), 220–38; Bagge *et al.* (1933), Part I, 400, 498, 561, Part II, 52–8.
7. Hirdman (1983), 20.
8. Archives of the Stockholm City Museum, folios 43 (Tobak), 54 (Stenarbetare, 85 (Klädsel), 122 (levnadshistoria), 127 (Tidningar och böcker). Workers seldom wore watches on the job, whereas foremen or supervisors wore them in order to maintain time discipline. Regarding new patterns of consumption in Sweden in general during this period see Szabo (1991).
9. Frykman and Löfgren (1979), 104–10; (1987), 194–205.
10. Miller (1987), 105.
11. Hall (1981), 26.
12. Cf. Williams (1982).
13. This is a position which rejects Baudrillard's (1983) extreme view of the consuming subject as merely a "victim," as simply an "uncritical recipient of signs and messages," as nothing but an "appendage to the commodity-object" (Tomlinson, 1990a, 20; cf. Ley and Olds, 1988; Bonnett, 1989). Regarding the cultural production associated with consumption see Miller (1987), Löfgren (1990a, 1990b), Willis (1990), and the literature cited therein.
14. Cf. Tomlinson (1990a).
15. Ahlberg (1958); Gustafson (1976).
16. Pred (1990a).
17. The Swedish counterparts of late Victorians are sometimes referred to as Oscarians, after King Oscar II, who reigned from 1872 to 1907.
18. Frykman and Löfgren (1979, 1985, 1987); Gaunt and Löfgren (1984).
19. Cf. Benedict (1983).
20. The term "society of the spectacle" was coined by Guy Debord (1983, 1990), a central figure of the avant-garde Situationist International, in his effort to

portray "how the meanings of power and ideology change as economics based upon mass consumption develop from within an economy of mass production" (Luke, 1989, 26). According to Debord (1983, no. 1): "In societies where modern conditions of production prevail, all of life presents itself as an intense accumlation of *spectacles*. Everything that was directly lived has moved away into a representation." Debord asserts that the "society of the spectacle," with its new forms of domination and alienation, had fully emerged by the 1930s.

21   Bennett (1988), 98, 84, 74, 76.
22   Allwood (1977); Benedict (1983); Rydell (1984).
23   Rydell (1984), 235.
24   Hughes (1991), 9.
25   Löfgren (1989a), 19.
26   Mitchell (1991), 149, 12–13, 6.
27   Davison (1982), 6.
28   The original proposal and surrounding debates are available in various parliamentary papers of 1895 (*Rikdagstryck*, 1895a, b, c, d). The considerable success of the Copenhagen Exhibition of 1888, among other things, had spurred Swedish manufacturers to maintain pressure for an exposition subsequent to the delaying action of 1890 (Hasselgren, 1897, 67). The fact that international exhibitions were to be held simultaneously in Brussels, Munich and Hamburg proved no deterrent in light of the regential anniversary.
29   *Allmänna Konst och Industriutställningen: Matrikel*, (1897); Hasselgren (1897), 82–4; Looström (1899b); Ekström (1989), (1991a), (1991b), 28. Here, as in several other instances later in this chapter, references are grouped at the end of a paragraph so as to avoid their unnecessary repetition.
30   Anderson (1983); Hobsbawm (1990), 91–3, 122.
31   Cf. Bennett (1988), 93.
32   The Swedish section of the Hall of the Arts contained no less than 667 works by 175 artists, most of which – in retrospect – have been deemed "banal drawing-room paintings." Critics, however, were generally deeply affected by the national posture of Anders Zorn, Ernst Josephson, Bruno Liljefors, Carl Larsson and other members of the "radical" *Konstnärsförbund*, or Artist's Union (Strömbom, 1965, 117).
33   Hasselgren (1897), 1,048.
34   *Luxemburger Zeitung*, August 29, 1897, as quoted in *Dagens Nyheter*, September 18, 1897.
35   An interviewed German cabinet minister, *Dagens Nyheter*, May 27, 1897.
36   An interviewed Danish exhibitor, *Dagens Nyheter*, June 9, 1987.
37   Excerpts from an article in *The Timber Trade Journal* as quoted in *Dagens Nyheter*, June 25, 1897.
38   Sjöström (1937), 71. As of 1895, close to 40,000 unmarried laborers slept on kitchen floors or other corners of space sublet from married working-class couples trying to make ends meet. Working-class living conditions in Stockholm during the 1890s are described at length in Geijerstam (1894),

Key-Åberg (1897a) and Tengdahl (1897).
39 The various educational objectives associated with the Stockholm Exhibition have been considered at length by Ekström (1991a). In addition to consumer education and the instillment of both national pride and a belief in the future based on continued technological progress, these objectives extended to improving the working capacity of industrial laborers and the provision of moral uplift.
40 Cf. Lundgren (1990).
41 Hasselgren (1897), 168.
42 *Nordens Expositionstidning*, September, 1896.
43 Hasselgren (1897), 1,051.
44 Tengdahl (1897), 16–17.
45 Lengthy accounts of the opening ceremonies, as well as the full text of the King's speech, are available in Hasselgren (1897, 164–72), Looström (1899b), and the Stockholm daily newspapers of May 17, 1897. Commentaries on the King's speech are available in Björck (1947, 354) and Ekström (1991b, 101–2).
46 *Fäderneslandet*, May 15, 1897.
47 Ekström (1991b), 98.
48 *Stockholms kommunalförvaltning* (1899), 172–3.
49 Key-Åberg (1897b), 50.
50 Hammarström (1970).
51 Bruno (1954), 143.
52 Reslow (1929), 101–2.
53 Larsson (1967), 17–18; *Stockholms Tidningen*, October 4, 1897; *Svenska Dagbladet*, October 5, 1897; *Nordens Expositionstidning*, August 13, 30, 1897.
54 Two-page cartoon in *Söndags-Nisse*, May 16, 1897. See Sörenson (1991, 1992) regarding the architecture of the exhibition, and especially the striking buildings created by Ferdinand Boberg, a key figure whose signature remains prominent on the landscape of central Stockholm.
55 Contemporary cited in Munthe (1948), 1,624.
56 See Celik (1992) on Islamic architecture at nineteenth-century European world's fairs.
57 The Hall of Industry was given its character by the imagination of Ferdinand Boberg who, in this instance, was assisted by Fredrik Lilljekvist, another leading architect of the period.
58 Strömbom (1965), 116. This building was also the work of Boberg.
59 Björck (1947), 356.
60 Olsson (1947).
61 Hasselgren (1897), 163.
62 International expositions in late ninetenth-century Europe generally did not have women's pavilions, although such events in New Orleans (1884–5), Chicago (1893) and Atlanta (1895) did have separate women's exhibition buildings that usually emphasized weaving and other productive activities

(Benedict, 1983, 40). A Women's Exhibition held in Copenhagen (1895) proved to be something of a fizzle.
63 Cf. the comparison of late nineteenth-century department stores and world's fairs in Lewis (1983).
64 Olsson (1947).
65 *Nordens Expositionstidning*, November, 1897.
66 Cf. Frykman and Löfgren (1979), 104–11; (1987), 194–205; Gaunt and Löfgren (1984), 40–4.
67 Hasselgren (1897); Folcker (1897); numerous newspaper reports, especially those appearing in *Nordens Expositionstidning*, May 18–November, 1897; photographic collection of the archives of the City of Stockholm Museum (*Stockholms Stadsmuseum*).
68 Buck-Morss (1986), 26.
69 This partial inventory derived from the same sources indicated for the Hall of Industry. Actually, the Machinery Hall was not 100 per cent devoted to producers' goods as washing machines, bicycles and a few other already mentioned household consumer goods were to be seen there.
70 Parenthetic sense mine rather than the author's.
71 Cf. Boyer (1992), 201.
72 Silverman (1977), 77.
73 Plentiful examples of almost all the objects indicated in this paragraph are to be found boxed among the archival holdings of the City of Stockholm Museum (*Stockholms Stadsmuseum*). The guidebook and maps used for orienting oneself at the exhibition also inevitably contained advertisements.
74 *Nordens Expositionstidning*, May 26, September 11, November, 1897; *Social Demokraten*, October 7, 1897; Hasselgren (1897), 1,035–9; Lindorm (1934), 347.
75 Björck (1947), 353.
76 Rosander (1989), 25, 32, 22.
77 Dalecarlia served this mythological image well because, unlike most other parts of the country, there were relatively few landless agricultural laborers amongst its population. However, some of the sterotypical qualities ascribed to Dalecarlian women were very likely in part derived from their performance as seasonal laborers in Stockholm, not only at cemetaries, but at breweries and construction sites that were bursting with labor-versus-capital tensions at the very moment of the Stockholm Exhibition.
78 Rosander (1989).
79 *Nordens Expositionstidning*, May 28, 1897.
80 Buck-Morss (1989), 81.
81 After elevators began operating in the two forward minarets on June 2 they were used by more than 1,000 people per day on the average (*Dagens Nyheter*, August 27, 1897; *Nordens Expositionstidning*, August 28, 1897).
82 Hasselgren (1897), 218–19.
83 Cf. Barthes (1979) on the Eiffel Tower, which was commissioned for the Paris

Exhibition of 1889. Also see Bennett's (1988, 97–8) discussion of Barthes.
84 Hughes (1991, 11), commenting on the Eiffel Tower as a concretization of how "the ruling classes of Europe" conceived the promise of technology.
85 *Dagens Nyheter*, September 30, 1897; *Nordens Expositionstidning*, June 22, September 30, 1897; cf. Ehrensvärd (1972), 27–33 and Johannesson (1978), 233.
86 Ekström (1991b), 108.
87 Hasselgren (1897), 250–2; *Nordens Expositionstidning*, September 6, 1897.
88 Jakobsson (1977); Matovic (1984).
89 Williams (1990), 8.
90 Cf. Williams (1990, 53–4, 153) with respect to the conflicting senses of fear and sympathy held by the well situated toward the lower classes. Also note other of her observations regarding the myths, narratives and ideologies associated with the underground in nineteenth-century Europe.
91 Williams (1990, 5), quoting Mumford (1934, 69–70).
92 Description of the Scanian Large-Scale Industry Pavilion and the "coal mine" from *Fäderneslandet*, June 23, 1897; and Hasselgren (1897), 922–5.
93 Hasselgren (1897), 361–2.
94 Williams (1990), 78.
95 Wallén (1971), 184; Hasselgren (1897), 223–6; *Illustrerad handbok*, 1897, 98–100; *Social Demokraten*, September 8, 1897.
96 Quote from a fictional figure in T. N. (1897). The attributes of the *Sagogrottan*, and the way in which it was experienced, are in keeping with Rosalind Williams' argument that gradually, between 1700 and 1900, Europeans came to perceive the underground not as ugly or sublime, but "as wholly beautiful, as magically illuminated artificial paradise, a splendid refuge from nature's imperfections" (Williams, 1990, 83). The illuminated, enchantment-suffused qualities of the *Sagogrottan* strikingly resemble those of the underworld paradise depicted in Gabriel Tarde's *Underground Man*, which was first published in 1896 (Williams, 1990, 102–3). The *Sagogrottan*, which was meant to be a money-making venture, also may be regarded as one precursor to the shopping malls and other late twentieth-century "pseudo-subterranean environments" in which "business exploits the imagination to market not just specific goods but a whole commodity intensive way of life" (Williams, 1990, 113).
97 Wallén (1971), 184; Hasselgren (1897), 583–7; *Dagens Nyheter*, June 10, August 21, 1897; *Nordens Expositionstidning*, June 18, 19, August 22, September 20, 25, 1897.
98 Stockholms kommunalförvaltning, 1899, 175–7, 315–17.
99 In a display at the Stockholm Pavilion, progress was didactically demonstrated by the city's Gas and Electricity Works by showing the history of public lighting since the use of tar torches.
100 Williams (1990), 103.
101 *Nordens Expositionstidning*, September, 1896.
102 At many international expositions foreign rather than domestic urban environ-

ments were recreated not only to create an image of the romanticized exotic, but also to underline the "civilizing mission" of Western culture, the backwardness, disorder and incapacity for change of colonized peoples and the Islamic world at large. Note Mitchell (1991) and Celik (1992) on the Egyptian and Ottomon Empire "quarters" of the 1867 Paris Exposition and the "Rue du Caire" of the 1889 and 1900 Paris expositions.

103   *Illustrerade familjetidskriften för svenska hem*, 1897, 51, 188.
104   Olsson (1947), 2.
105   Hasselgren (1897), 104.
106   Wallén (1971), 184.
107   Here, as elsewhere in *Gamla Stockholm*, anachronistic spelling was used as an element in creating the illusion of a living past.
108   Hasselgren (1897), 304.
109   The Rue du Caire at the 1889 Paris Exhibition similarly abounded with opportunities for consumption and paid entertainment. One could pay for perfumes, pastries, tarbushes, donkey rides and dance performances among other things (Mitchell, 1991, 1).
110   Aside from the cited quotes, the depictions of this section are derived from: *Dagens Nyheter*, May 21, June 16, 1897; *Fäderneslandet*, June 2, 5, 1897; *Nordens Expositionstidning*, June 5, 15, 18, 29, July 20, 21, August 22, 25, 31, September 6, 8, 20, 23, 1897; *Social Demokraten*, July 6, 9, 1897; Cornell (1952), 241; Folcker (1899); *Illustrerade familjtidskriften för Svenska Hem*, 1897, 188–90; Idestam-Almqvist (1948), 2,250–1; Lindorm (1934), 358; and Hasselgren (1897), 292–315, 568–9. Cf. Ekström (1991b) 114–17 and, by way of parallel, Löfgren's discusssion (1989b) of the bourgeois idealization and nationalization of the Swedish natural landscape during the nineteenth century. Numerous other ideologically-laden spaces of illusion existed in the dream-world of the 1897 Stockholm Exhibition, not the least of which was the "Naval Spectacle" in which thirteen electrically driven miniaturized vessels did gunfire and torpedo battle with one another in the adjacent inlet waters.
111   Benjamin (1976), 165.
112   *Söndags-Nisse*, July 4, 1897.
113   *Idun*, December 17, 1897; *Svenska Dagbladet*, October 7, 1897; *Nordens Expositionstidning*, June 15, 21, 28, July 23, September 11, 13, 1897; Blanche (1897); Hasselgren (1897); Söderberg and Rittsel (1983), 53; plus souvenir objects held in the archives of the City of Stockholm Museum (*Stockholms Stadsmuseum*).
114   Calinescu (1977), 223–68; (1978).
115   *Dagens Nyheter*, January 29, April 9, 15, May 17, June 2, July 6, 1897; *Nordens Expositionstidning*, May 28, 1897.
116   Allwood (1977, 180) puts the total attendance at the 1867 *Exposition Universelle* at a considerably lower 6.8 million.
117   Allwood (1977), 181–2.
118   Cf. Ohlsson (1959), 194.

119 *Fäderneslandet*, June 2, 1897.
120 Bendix (1899); *Dagens Nyheter*, August 23, 28, September 25, 1897; *Social Demokraten*, July 13, 1897.
121 Looström (1900), 1,072.
122 Bendix (1899), 114; *Dagens Nyheter*, February 19, 24, April 7, June 5, 8, 14, 1897; *Nordens Expositionstidning*, June 1, 9, 1897.
123 The exhibition administration also pursuaded the State railways temporarily to provide roundtrip tickets to Stockholm for the price of a one-way ticket (Blanche, 1897, 36). However, these reduced fares, which were available from every station in the country, remained prohibitively expensive for most industrial and agricultural wage-earners, and thereby were primarily to the advantage of the bourgeoisie. Discount fares, affordable to working-class members were also arranged on "People's Steamboats" from northern ports on the Gulf of Bothnia (*Nordens Expositionstidning*, August 5, 10, 14, 1897; *Dagens Nyheter*, October 1, 1897; *Social Demokraten*, August 13, 1897).
124 *Nordens Expositionstidning*, July 2, 1897; Molander (1899), 149.
125 *Social Demokraten*, July 2, 1897.
126 A total of 96,165 people were admitted to the exhibition by way of block tickets, only ten of which had to be used at any given time (Bendix, 1899, 113). As such tickets were also eventually provided to groups of secondary school students and locally garrisoned troops, the entire total cannot be ascribed to industrial employees. However, as the frequency with which the press referred to specific worker-group visits (and their numbers) was many times greater than the frequency with which it referred to specific school-group visits, the estimate of 70,000 may well be conservative. In addition, some Stockholm manufacturers supplied their laborers with full-season passes. Office workers and retail employees, especially from within Stockholm, were also frequently given free visits to the exhibition by their employers (who were not qualified for block-ticket purchases).
127 *Dagens Nyheter*, June 16–September 22, 1897; *Nordens Expositionstidning*, June 16–October 1, 1897; *Social Demokraten*, July 1–October 6, 1897. Many farmhands on smaller farms around Stockholm requested an extra day off in order to visit the exhibition at their own expense (*Dagens Nyheter*, June 26, 1897).
128 Pred (1990a), 129–30.
129 Frykman and Löfgren (1985), 129.
130 Because they arrived in an exhausted state, because they had only one day in which to make use of their free tickets to *Gamla Stockholm* and five other fee-charging features, and because of the time they wasted waiting for meals at the "People's Kitchen" (see above), many People's Train visitors must have often experienced a time-pressure and discipline akin to that they normally experienced at the workplace (*Dagens Nyheter*, July 7, 1897; *Nordens Expositionstidning*, July 13, August 3, 1897; *Social Demokraten*, July 16, 1897).
131 Bendix (1899), 114; Ekström (1991a); Molander (1899), 152; *Dagens Nyheter*, June 2, 17, 28, August 2, 1897; *Fäderneslandet*, May 12, 1897; *Nordens*

Expositionstidning, July 2, 9, 12, 19, 22, 1897; *Social Demokraten*, July 2, 3, 8, September 21, 1897.
132 Bendix (1899), 110, 117–21; Blanche (1897), 42; *Dagens Nyheter*, May 31, 1897; *Fäderneslandet*, May 26, June 2, 1897.
133 *Nordens Expositionstidning*, July 31, 1897.
134 Blanche (1897), 72; Hasselgren (1897), 752–3; Zethelius (1899); *Dagens Nyheter*, July 23, 1897; *Illustrerade familjetidskriften för svenska hem*, 1897, 115; *Nordens Expositionstidning*, photographic collection of the Stockholm City Museum (*Stockholms Stadsmuseum*).
135 Rydell (1984), 2.
136 Rydell (1984), 2, based on Berger and Luckmann (1967).
137 Hasselgren (1897), 1,041–2.
138 *Dagens Nyheter*, October 4, 5, 1897; *Fäderneslandet*, October 6, 9, 1897; *Nordens Expositionstidning*, November, 1897; *Nya Dagligt Allehanda*, October 4, 1897; *Social Demokraten*, October 4, 1897; *Stockholms Dagblad*, October 4, 1897; *Stockholms Tidningen*, October 6, 9, 1897; *Svenska Dagbladet*, October 6, 1897.
139 *Nordens Expositionstidning*, October 2, 1897.
140 *Söndags-Nisse*, October 10, 1897.
141 Rudberg (1986), 118.
142 M. Löfgren (1991).
143 Benjamin (1972–89), V, 115, as quoted in Buck-Morss (1989), 99.
144 Cf. Buck-Morss (1989), 283–4.
145 Cf. Miller (1987, 70) on Simmel, consumption-good objects not of one's own making, and desire.
146 *Stockholms Tidningen*, October 9, 1897.
147 *Social Demokraten*, October 9, 1897.
148 *Nordens Expositionstidning*, September 6, 1897.
149 Cf. Frykman and Löfgren (1985), 122.
150 Bendix (1899), 116; *Dagens Nyheter*, July 23, 1897; *Nordens Expositionstidning*, June 30, July 10, 14, 22, 23, 27, 31, August 9, 12, 18, 20, 24, 30, September 17, 30 1897; *Stockholms Tidningen*, October 7, 9, 1897.
151 *Svenska Dagbladet*, October 4, 1897.
152 *Social Demokraten*, October 4, 1897.
153 *Social Demokraten*, October 6, 1897.
154 *Social Demokraten*, July 13, 1897.
155 *Nordens Expositionstidning*, September 24, 1897.
156 Hasselgren (1897), 1,042.
157 *Nordens Expositionstidning*, November, 1897.
158 Hasselgren (1897), 1,042, 1,044; *Nordens Expositionstidning*, November, 1897.
159 Looström (1900), 1,072.
160 Buck-Morss (1989), 286.
161 Benjamin (1972–89), V, 130, as quoted in Buck-Morss (1989), 286.
162 Uncataloged item dating from 1901, held at the archives of the Stockholm City Museum (*Stockholms Stadsmuseum*).

CHAPTER TWO

PURE AND SIMPLE LINES, FUTURE LINES OF VISION: THE STOCKHOLM EXHIBITION OF 1930

•

PROLOGUES

A byproduct of this [post-war European mood] was a sense of transitoriness. Whether in fashion, architecture, or the tableaux of Piet Mondrian, curves were abandoned in favor of straight lines, lines that suggested movement, a new simplicity, and a new beginning.	MODRIS EKSTEINS (1990), 259
More than any other date since the industrial revolution [in Sweden], 1930 constitutes a boundary line between old and new.	GÖRAN THERBORN (1981), 25–6
For no one in Europe (or the United States) could have lived through the decade of the thirties without being aware that international expositions, having become less frequent after World War I, suddenly came back with a vengeance during the Depression years. They were seen as a means of enhancing business, creating jobs for the unemployed, and providing state-subsidized, mass entertainment that was at the same time public "education." Major expositions occurred almost yearly: Stockholm 1930;... New York 1939.	SUSAN BUCK-MORSS (1989), 323[1]
The [wave of] social [as opposed to industrial] modernization was launched with thunder and lightning by the Stockholm Exhibition of 1930.	LORENTZ LYTTKENS (1991), 18
The idea of the Stockholm Exhibition, so far as I could understand it, originated in, was derived from, and led into a social upheaval encompassing all of society, towards a bloodless revolution. The exhibition expressed both a	EYVIND JOHNSON (1961), 28

new conception of the beautiful and a high social ethic: build for people in a society that is in the process of becoming better, that soon will become better.

PER G. RÅBERG
(1980), 9

[F]or many, and especially the young generation, the Stockholm Exhibition of 1930 was first and last a promise-rich vision of the city of the future, of the dreamed of technological society.... It was a moment of boundless confidence in the ability of technology and rational reason to create a better society and a more beautiful world.

MAX OSBORN
in *Vossiche Zeitung*, as quoted in *Svenka Dagbladet*, August 15, 1930

The praiseworthiness and significance of the Stockholm Exhibition lies in the bold grandiosity with which it has adopted and animated a new epoch's way of thinking in order to create a document of the times sparkling with inner life. Here, moreover, the international victory of modern design thinking is confirmed.

GILLIAN NAYLOR
(1990), 164

For P. Morton Shand in the [August 1930 issue of] Architectural Review, the Stockholm Exhibition demonstrated a complete and triumphant break with the past; it also demonstrated that Sweden was the only European country capable of producing a viable form of Modernism. According to Shand, "Sweden could do it better than the Germans and the French, and most certainly the English: The world will look up to Sweden," he wrote, "as the supreme exponent of a Modernism which has succeeded in finding its own soul and embellishing itself with a purely mechanistic grace."

J.M. RICHARDS
(1940), as quoted in Gillian Naylor (1990), 179

[I]t was in Scandinavia that an event took place ... which did more than anything else to arouse public interest in modern architecture: the Stockholm Exhibition of 1930 ... for which the architect was Gunnar Asplund. Previously modern buildings had been seen only in the form of isolated structures that inevitably looked stranger than they really were when surrounded by the mixed architectural styles of the average city street, but at Stockholm a whole sequence of buildings – as might be a whole new quarter of a town – were designed and laid out in a consistently modern style, and the public, walking among them, was given a first glimpse of modern architecture not as a new fashion in design but as a newly conceived environment.

M. KYLHAMMAR
(1990) as quoted in Lorentz Lyttkens (1991), 18

The Exhibition was regarded in fact as a demonstration of what a modern rational existence meant: household appliances, mass produced "machine houses," "functional" furniture and, not to be forgotten, the dream of the new [hu]man [being] in the just society.

No frills. Only orderliness. That is an order. For the modern future.

Directly to the point. Right from the start. This account of the Stockholm Exhibition of 1930 is, after all, a set of (geographical hi)stories about rationality and unwavering certainty. A multi-voiced, many-stranded tale of designing social engineers whose no-nonsense-accepted mode of articulation was the pure and simple line, the spare geometric form. The pure and simple line. Without ornamentation. Without ambiguity. The pure and simple line as the signifier of a national modernization project. The pure and simple line as a line of vision into the future. An envisioned future where "progress," a new aesthetic of consumption, and the construction – or social engineering – of a new Swedish (hu)man are superimposed, conjoined and collapse into one another. Into a single point that moves relentlessly forward in step with the interests of both the reformist labor movement associated with the Social Democrats (who soon were to assume power) and internationally successful Swedish capital.[2] The undecorated, nothing-but-the-object(ive)-facts pure and simple line. The pure and simple line as a doubt-free boundary line, as encloser of a unitary no-difference-allowed space. On the inside of the line is a new sense of national identity, a new sense of solidarity, a new sense of achievement, the sensibly modern, the rational and enlightened individual who bears her expanded freedom "with a large portion of social responsibility."[3] Without question! On the other side of the line, outside and beyond, is the different, the inferior, the repudiated, the disapproved and condemned, the unmodern, the irrational, the unenlightened. Without question!

The unadorned, uncompromising, austere, pure and simple line. The pure and simple line, starting here and now, as a break with the past. The pure and simple line as new liberated shape, as liberating shaper, unanchored from the historical deadweight of Oscarian (Victorian) bourgeois Kultur, free from the superfluous squibbles and curlicues of conservative bourgeouis form.

The stripped bare, thrusting straight on pure and simple line. The pure and simple line as erection of masculine rationality, as principal signifier of a new phallocentric universe. The pure and simple line as authorized, authoritarian word, as rationality for a rationally controlled world. The pure and simple line as delivered word, as a carefully mapped new social trajectory. The pure and simple line as a planned path to cosmopolitan high modernity, as a promising, promised road to the better. A pure and simple line intended to change the world in desired ways.

A pure and simple line all the same leading to long-term unintended consequences, contributing to an unintended Swedish present. A present where the crisis of the welfare state, where the attendant loss of faith in progress-ive social engineering, where the constantly shifting investments of domestic and international capital, and where the boundless consumption heaven promised by market- and Common Market-worshipping "new liberal" politicians are all synonymous with a disorienting disjunction between long-standing and new meanings, between once unproblematic and now multiple, shifting meanings. A present where not only the meanings of geographically specific details of everyday life, but also the meanings of "national," "neutral," and "social democratic" can no longer be taken for granted, can no longer be experienced as second nature, can no longer be regarded as common sense, or idea-logical. A present, consequentially, where a variety of deep individual and collective identity crises abound.[4] A hypermodern, modernity-speeded-up present whose unmoored identities and cacophony of meanings in no small measure involve a cultural reworking of the pure-and-simple-lined idea-logics first given a spectacular articulation in Sweden at the Stockholm Exhibition of 1930, or the Stockholm Exhibition of Arts, Crafts and Home Industries, as it was officially known.[5]

In fundamental ways the Stockholm Exhibition of 1930 resembled past international expositions and "world's fairs" held in Europe and North America. Here was the summer-long spectacle as national project. Here a program for progress was made concrete. Here a politically charged ideology was given material expression. Here a universe of new consumption possibilities was given spectacular articulation, was presented in concentrated form, was displayed in an extraordinary time and space, was exposed to the senses in a time and space isolated from the routines and project demands of the everyday, was put on exhibition to be eyed and experienced under circumstances free from everyday responsibility. All the same the Stockholm Exhibition of 1930 was not an international exposition in the conventional sense of the term. No foreign country had a pavilion or display of its own there. No country other than Sweden was there to peddle its self-image. No foreign country came to do economic battle by way of puting its manufactured goods on view. No foreign country came to do political battle by way of showing off puffed up national symbols. The 1930 Stockholm Exhibition *was* international to the extent it involved the translation of international forms of modernity into a monolithic Swedish form. Into a single pure-and-simple-line form. The 1930 Stockholm Exhibition also was international to the extent it was intended to promote Swedish design in other countries, to expose certain Swedish goods to foreign critics and wholesale purchasers, to market Sweden abroad. Quite simply.

Practically every person is interested in those manufactured goods by which our homes are formed, and therefore goes to an industrial arts exhibition with a completely different and more spontaneous interest than to a pure industrial exhibition, which generally requires expertise in order to be properly understood.

From a petition to the crown regarding the exhibition (Kommittén för utredning av allmän svensk utställning ... [1928a], 6–7)

The Stockholm Exhibition of 1930 came into being as the result of an initiative undertaken by the Swedish Handicrafts Association (*Svenska slöjdföreningen*), an organization long devoted to the promotion of Sweden's arts and crafts and home furnishing industries, not least of all by the means of both large and more modestly scaled exhibitions.[6] Initially encouraged by a resounding success at the international arts exposition held in Paris during 1925,[7] and later swelled by the prestige of a major exhibit of glassware and other items at New York's Metropolitan Museum of Art,[8] the association – led by its determined chief executive, the art historian Gregor Paulsson – appointed an exploratory committee and then, with the support of the Swedish Domestic Handicrafts Sales Association (*Svenska hemslöjds försjälnings föreningen*) and important manufacturers, began seeking financial support from state, municipal and private sources.[9] In petitioning the Crown for permission to raise funds by way of state lottery drawings, it was essentially argued that the exhibition was in the national interest: while the leading "civilized countries of Europe" are each experiencing a "purposeful ... movement for the expansion and improvement of the industrial arts," it still can be said that "in no other country have the industrial arts ... attained such vigorous development as in Sweden."[10] And yet, a "peculiar set of circumstances" prevails, for "our modern industrial arts up to now have not been exhibited to any great extent in our own country."[11] Nothing on a grand scale has occurred domestically since 1909, a date when there was not even a trace of current designing trends. Surely the time is ripe to reaffirm Sweden's leading international position, to "awaken the entire country's attention; to give our industrial arts the domestic fame 'they deserve.'"[12] The summer of 1930 would appear especially appropriate since a national agricultural fair is already planned for a space adjacent to the waterside site proposed for the exhibition – a strikingly attractive site in the northern Djurgården area just beyond the city's highly fashionable embassy district; a site that will be made readily accessible by the laying out of a landscaped promenade, the construction of a special trolley line, the provision of a new bus line and two steam-launch services and the availability of a parking lot capable of accommodating 2,800 cars.[13] "It is as clear as day that it would be of the greatest value to expose products of the arts and crafts and home furnishing industries to the crowds

of rural residents of different strata who are expected to appear."[14] And, of course, the city of Stockholm will reap "enormous direct and indirect benefits" from the exhibition.[15]

The existence and objectives of the exhibition obtained a special legitimation through the Crown's appointment of its administrative board, its executive committee, and its general commissioner following another petition from the Handicrafts Association.[16] Here were organizational units whose honorary chairman was no one less than His Majesty the Crown Prince, whose actual chairman was no one less than the Minister of Justice. Here were organizational units whose membership included the rich and the famous, the prestigious and the powerful. Here, as with elsewhere-held international expostions and world's fairs, was a space and an event whose governance was a matter reserved for the élite.[17] Here sat Marcus and Knut H. Wallenberg, leaders of the country's pre-eminent financial dynasty. Here sat three other bank presidents, several corporate executives, department store owners, publishers, members of the aristocracy, various political office holders, the chairman of the Swedish Confederation of Trade Unions, and distinguished academics, as well as prominent architects, designers and cultural personages. Here sat Gregor Paulsson exactly where he wanted to be, in the driver's seat, now wrapped in the mantle of general commissioner, surrounded by all of these establishment figures, still able to function as the exhibition's principal driving force and chief ideologue. Here sat Paulsson and the twelve-man Executive Committee with legitimated power concentrated in their hands, with nothing below them but thirteen small hand-picked advisory groups: a construction committee; nine "expert" committees and review panels responsible for some of the more important product types to be displayed; and three special-event committees.[18] (In contrast, The Central Committee for the Stockholm Exhibition of 1897 had an administrative colossus of roughly 175 committees and subcommittees beneath it.[19]) Here sat Paulsson and his personally selected chief architect, Gunnar Asplund, able to control every aspect of the outward appearance and context of the exposition; able to determine which compromises – if any – were to be made; able to author-ize every exhibition feature; able to control even the most minute of details associated with each of the three exhibitionary divisions (architecture and construction materials [the home of the new-times "man"], streets and gardens, plus "means of communication" [the exterior environment of that "man"];[20] household goods and furnishings [the interior of "his" home]); able to control the twenty-five classes of objects sorted under these divisions; able to control the content of, and solely license, the distribution of all postcards, picture albums, souvenirs and popular songs pertaining to the exhibition;[21] able to determine specific regulations pertaining to exhibitors and their employees as well as exhibition visitors.[22] Here – within this spectacular space – legitimated authority would not permit, would not tolerate, any deviation from its design-ated pure and simple line. Here there would be nothing displayed but objects of contemporary Swedish manufacture, nothing but objects of "high quality in terms of form and design." Here – except for some traditional handicraft items – there would be nothing but the high(ly) modern. Here NO copying of older styles. NO imitation of older objects.[23] Here NOthing but the forbidding pure and simple line.

And what, more precisely, were the stated economic and social objectives? What were the articulations of modernity giving direction to the exhibition? What was the advance(d) word for public consumption? What was the lofty rhetorical line? What were the differences to be made?

If "properly managed" the exhibition would serve as both an educational and an "advertising form of great significance."[24] It would "raise the taste and cultivation of our entire population."[25] Contemporary, well-designed Swedish products, usually – but not always – resulting from the collaboration between artists and manufacturing firms would be made "better known for the Swedish public at large."[26] Although some expensive items would be placed on view for the luxury consumer, it was principally the "broad masses" who would be targeted, who would be relieved of their ignorance, who would be exposed to "more beautiful everyday goods" that were also functional. It was principally the "broad masses" who would be made subject to more aesthetically pleasing goods, to goods that wed "appealing appearance" with practicality and quality.[27] By taking advantage of the fact that home decoration has become central to "the private interests of people" in new ways, "the sales possibilities" of the exhibiting firms would be increased.[28] The competitive atmosphere of the exhibition would encourage participant manufacturers "to develop and perfect their products further."[29] The quality of "socially significant" production would be improved and its profitability thereby ensured.[30] Mass production and mass consumption would assume new forms. "Good" and "practical" items, produced in large series, would become available to the "great consuming masses."[31] Mass production and mass consumption would be brought more fully in line with the times, would catch up with the times. Mass production and mass consumption would be brought more fully in line with the "structural changes which have occurred within society's technological and social realms, [and that] are in the process of creating a spirit of the times, a sense of life, or whatever you might call it, that objective observers consider as different from the foregoing as, for example, the Renaissance from the Middle Ages."[32] By means of its "propaganda" function the exhibition would support manufacturers who strive toward "the common goal" of using product design to "give the environment in which we live a dignified character and quality as well as a form in keeping with the leading features of the [new] times."[33] By bringing newly designed products and newly designed housing before the eyes, by doing so in a setting devoid of "museum-like solemnity," it would be demonstrated "that [these items] are an urgent matter for every inhabitant of our country."[34] Not least of all, the exhibition would confront Sweden's housing problems, would address housing problems that were most pronounced in Stockholm, "would attempt to create the best possible dwelling units for given household types and given incomes," "would strive for the *purest* possible solutions," and therefore "would not take any regard of solutions currently in vogue."[35] Through a cluster of model apartments and detached residences it would be shown that it is possible to provide affordable housing which takes "the practical realities of everyday life" into account,[36] which is not substandard, which makes rational use of limited floor space, which is social(ly) democratic. Through its presentation of new homes for a new world, through its product and housing displays, the exhibition would contribute to "the formation of a new high-quality society" and a new urban culture, one which turns its back on the long-dominant culture "of country roads, peasant villages and provincial towns."[37] At this dramatic juncture in time, with the help of the exhibition, a "radically new style" and aesthetic

HOUSING CONDITIONS IN THE CITY OF STOCKHOLM, 1930

Against the background of the often miserable housing conditions of the working class [in Stockholm], housing policy had become the great topic of [local political] debate during the postwar period, with high levels of tension between the parties, irritation-laden discussions, and heated exchanges.   Yngve Larsson (1977)

Population growth in greater Stockholm 1921–30: 113,800, of which in here-mapped central city 52,400 (an increase of 23.3 per cent).

Central city conditions at end of 1930:
48.2 per cent of all dwelling units consisted of either 1 room plus kitchen or kitchenette, or 1 room without cooking facilities
112.6 residents per 100 rooms
48,267 people, or 11.5 per cent of the population, dwelled in "overcrowded apartments" (more than 2.0 persons per room)
65.7 per cent of all dwelling units did not include a bathtub or shower
59.0 per cent of all buildings had no central heating

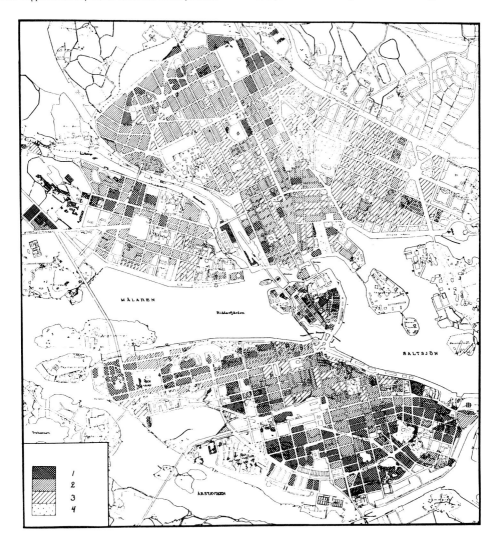

Residential density in street-abutting buildings, Stockholm, 1930.
Density scale (rooms per resident): (1) < 0.8; (2) 0.8–1.0; (3) 1.1–1.3; (4) > 1.3.

Despite considerable construction during the 1920s, real wage increases and decreased nativity, the working-class population of Stockholm was still largely confined to dwelling units that were pitifully small and poorly heated, lit and ventilated. Low vacancy rates and high loan charges usually helped make these units rather expensive, even for those taking in boarders or receiving some kind of subsidy from municipal authorities. Monthly housing expenses were often "flagrantly disproportionate" to income. All in all, housing conditions in Stockholm, and Sweden as a whole, were significantly worse than any other industrialized European country.

In the Stockholm municipal election of 1931, the Social Democrats for the first time achieved a majority of their own, having campaigned in a welfare program of ten points, foremost of which was "healthy and inexpensive housing."

Sources: Ahlmann *et al.* (1934), 55 (map); Larsson (1977), vol. 1, 235, vol. 2, 388 (extended quote); Johansson (1942), 172–4, 231–2; Myrdal and Myrdal (1934); Stockholm stads statistiska kontor (1930), 100 and (1931) 92–3; Svedberg (1980).

would triumph.[38] The pure and simple idea-logical line of "functionalism" would triumph. The architects of modernity would triumph. The new architect, the architect as clearly thinking, socially oriented technician and organizer would triumph. People would come to recognize that "the form of a household good or building should be as faithful an expression as possible for the object's function."[39] People would come to recognize that "the practical is the beautiful."[40] The mode(rn) would be redefined. Linear orderliness! High time!

But who would buy all or any of this?

Who would come to identify with the new housing, the new household furnishings, the new prescription for urban design and urban living?

Who would buy into a program that sought to modernize consumption and private life, that served the interests of industrial capital and yet was congruent with the Social Democrats' emerging imagery of a "People's Home," of a big home for one and all, of a society peopled by rational, enlightened and socially responsible citizens, of a society which was to provide a metaphorical shelter for the weak and concrete functionalist shelters for the now poorly housed?[41]

Who among the well-to-do, among the "broad masses," among the urban population of Stockholm and lesser cities, among the country's rural residents, would absorb these new messages, would allow their meanings to become central to their identities?

Where were the women and men who would immerse themselves in this "program for reschooling Swedes into modern consumers by letting things teach them the lessons of modernity"?[42]

Gregor Paulsson: "The arrangement of an exhibition is a piece of Roman statescraft [carried out] within the realm of commerce. It is not merely a matter of exhibition goods; one must also see to it that people come. And the people demand both bread and circuses."[43]

It was recognized that it would require something more than reduced price admission days if the objectives of the exhibition were to be met; if the exhibition's architects and product engineers were to be successful social engineers; if they actually were to contribute to the emergence of the new society, the new culture, the new person; if large numbers of people were to be attracted straightaway to come and see the straight way, the pure-and-simple-lined way. It would take bread and circuses to draw people. And bread and circuses they were given. At the "Festival Plaza" facing the principal

THE STOCKHOLM EXHIBITION OF 1930

Gouache, 1925

exhibition buildings they were given one special event after another, usually of massive proportions. The performance of two grandly scaled theatrical shows with voices projected from Europe's largest public-address system. A fly-over by the Graf Zeppelin. A folk-dance festival with 1,500 participants. A demonstration of mass gymnastics. An ambitious garden show. And always music, more music. A conglomerate of choruses singing together in conjunction with a Nordic Choral Festival. Daily concerts by a military band and the Exhibition Orchestra (members of the Stockholm Philharmonic). The strains of Berény's Hungarian Gypsy Orchestra.

At the waterfront and the "swimming stadium" adjacent to the Festival Plaza they were given motorboat races, rowing regattas, eye-dazing bedazzling fireworks, the water and light spectacle of *La Fontaine Lumineuse*, swimming competitions, diving contests, water polo matches between the Swedish and Hungarian national teams.

At the amusement park at an eastern extreme – where its structures would not interfere with the harmony of the remainder of the exhibition area – they were given a heavy dose of jingle-jangle and diverse entertainments. Given real circus acts: The Three Aerial Flacks; The Four Flying Beauties; a troop of Cossack horsemen. Given "machine romanticism ... sensations under the sign of technology."[44] Given the opportunity to ride a mini-zeppelin (until its demise on a trip back to Germany for repairs), to experience the ups and downs of a roller-coaster and a Ferris wheel, to play whirl and bump with "radio"-powered mini-cars. Given "Europe's largest" open-air dance floor with space for 1,500 couples to swing to the music of Jack Stanley and his Florida Boys (all Swedes). Given another dance area where the older could shuffle to more sedate tones. Given performing people as items of consumption: the Musical Review Theater starring siren Zarah Leander (later the darling of Germany's Nazi leaders) and the ten dancing Funkis Girls; the Magic Theater "where Max and Alva v. Rhodin carry out amazing experiments in modern magic;" the Specialty Theater, where a constantly changing array of continental sensations appeared. Given standard carnival fare such as a hall of mirrors, shooting and archery galleries, pinball machines, countless knock-over-the-object and spin-the-wheel games. Given a Punch and Judy show and a separate "Children's Tivoli," a merry-go-round, tame animals, storytellers, a juice and waffle bar. Given food and drink alternatives at four restaurants and tea-room cafés as well as by wandering vendors. (Bread rather than circuses also was provided at nine food and drink establishments within the exhibition area proper.)[45]

Between May 16 and October 9, 1930, over 4.1 million people visited the Stockholm Exhibition of Arts, Crafts, and Home Industries.[46] From the very outset the functionalist exhibition as a whole, as well as its pure and simple architectural and interior-design style, were popularly labelled *"Funkis."*[47]

Was this just another Stockholm slang abbreviation, with its characteristic -*is* ending?[48]

Or was this a symbolic bringing down to earth of the exhibition and its everywhere-to-be-seen official emblem, a pair of distinctive wings, the wings of progress?

How many among the attending millions were buying any or all of the exhibition's high-flying rationalities and offerings of modernity?

How many were coming only for bread and circuses?

How many were coming after 7 p.m., when the cost of admission was cut in half, when the amusement park was in full operation, but when the exhibition halls and pavilions were closed?

How many were identifying with the new housing, the new household furnishings, the new prescriptions for urban design and urban living?

How many among those differenc-iated millions were absorbing the messages offered, were allowing the meanings offered to become central to their identities?

Take a partial look. And another. And another. Take a simultaneous voice-listen and construct your own reading, construct your own (geographical hi)stories of the high modern.

MONOLOGUES, DIALOGUES, POLYLOGUES, IDEOLOGUES

I walked along the main street of the big 1930 Stockholm Exhibition. It was summer and stifling hot. The sun of the new decade shone on my crown. An entire new city of steel, glass and cement had been erected on the flat area where before there had been a void. Houses, restaurants and music grandstands which resembled birds flying upward with stiff wings. Among the crowd people spoke of the new architecture which would give birth to a new sense of life. A door handle, a picture window, a matter-of-fact piece of furniture would in a short time influence the family residing in a house in such a way that their feelings and thoughts would become open, transparently clear. The shining machine limbs of the exhibition halls demanded a new poetry. The exhibition area's high steel mast rose like a signal, like a shiver of happiness toward the bright blue sky. The functionalist era had blown in. The style of the new age was really the scraping away of styles. Its naked language is called facts. I directly translated the language of architecture to that of literature. I walked and looked about for the new [hu]man [being].

"Proletarian author" and journalist IVAR LO-JOHANSSON (1957), 5

Many still remember . . . the exhibition's airy white pavilions and festively flapping flags.

HENRIK O. ANDERSSON (1980), 3

ALVAR AALTO
as interviewed in *Åbo Underrättelser* and quoted in Kenneth Frampton (1985), 37

The biased social manifestation which the Stockholm Exhibition wants to be has been clad in an architectural language of pure and unconstrained joy. There is a festive refinement but also a childish lack of constraint to the whole. Asplund's architecture explodes all the boundaries.

PER G. RÅBERG
(1980), 9

It is difficult to imagine a more seductive argument for the rationalism that was breaking through than Gunnar Asplund's supreme exhibition architecture.

IVAR LO-JOHNSSON
(1979), 452–3

Everything that might obscure was taken away, every bit of rubbish that might prevent the observer from seeing into the future was removed. It was a city almost built on air. Or it was like a ship built of steel and glass with the main street, or Corso, as deck and the 75-meter tall tower as mast. Any and every hint of ornamentation was forbidden. . . . It was a city without falsehood. The main building was reflected in an artifical pond which laid shining as if the water-surface was also artificial. Transparent grandstands sort of hovered. Flags lashed and snapped in the wind. The whole thing felt as if one was on a street which leads straight into the future. One was already in Urbs, The City of the new [hu]Man [being].

Within the exhibition area there was everything one could demand of a city. There were restaurants, a telegraph station, a post office, a press bureau, a hotel agency, a planetarium . . . as well as . . . an aquarium. . . . There was a movie theater [which showed advertising shorts], . . . , a police station with a clink, a fire station, a sanitary works, public toilets, administrative buildings, an infirmary, but not a home for the aged.

Exhibition area dialogue in
EYVIND JOHNSON
(1948), 748

"Wonderful, one is crushed by it all. And God, how tired one's feet get."

"It's like a revolution, so quickly it's grown out of the ground."

"It's a beginning, the obvious beginning, I mean; the obvious, clear beginning to a new and better world."

"[The architects] have discovered how people ought to live, how society ought to look."

"The times are on the rise in Norden, now. Yes, in the whole world."

Social Demokraten, May 16, 1930

Condensation of a rhyming cartoon caption: You can say what you want about the exhibition city's architecture. It sure will create jobs for window washers.

Ny Dag (communist daily), May 16, 1930

Swedish capitalism's propaganda exhibition of 1930 was ceremoniously opened today. A grandiose and successful advertisement for "Swedish enterprise" – a delight for the eye.

Nothing new here. This all reminds me of things I saw in Moscow in the early twenties.

Extracted, paraphrased essence of an interview with Albert Engström in a May, 1930, number of *Svenska Dagbladet*

Gregor Paulsson: "Do you think the exhibition is predominantly international or national in tone?"
Foreign visitor: "It's international but with a Swedish aroma; one might say Le Corbusier with a shot-glass of aquavit."

Social Demokraten, May 25, 1930

I can't recall any exhibition where after a tour I felt refreshed rather than tired. But such is the case with the Stockholm Exhibition of 1930 to a surprisingly high degree.

A New York journalist as quoted in *Stockholms Tidningen*, June 7, 1930

[T]he new city [of the exhibition area] was formed not so much in steel, glass, and cement as with 60,000 square meters of "Eternit" sheets.[49]

LEIF NYLÉN (1991)

Men saw the exhibition in their own way, and women in another.

IVAR LO-JOHANSSON (1979), 455

If Le Corbusier had said that a house was "a machine to live in," then this exhibition was a machine to hold a festival [or party] in.

GUSTAF NÄSSTRÖM (1961), 43

If one assembles a group of photographs [from the Stockholm Exhibition of 1897] and compares them with the sketches for the upcoming exhibition, one can't avoid being struck with how great the difference is between these two expositions, situated on opposite sides of an inlet, only a few hundred meters wide. It is an ocean that separates them.

GOTTHARD JOHANSSON (1929)

 The approaching exhibition is namely built according to a consistently drawn up plan, it has strict objectives and purposes, and its architecture is meant to work as propaganda and model. The 1897 exhibition had no such intentions. It was in its way a fairy-tale city, where all kinds of styles and whims were accepted; it offered a few main buildings and a lot of pavilions, which each proprietor was permitted to dress up according to his own taste....

 ... 1897 was a romantic construction, a free fantasy, a color-sparkling dream city if one so will.... The 1930 exhibition wants to be "prose, the reality of life," as Tegnér says.

If to enter the Stockholm Exhibition of 1897 – or any other typical international exposition – was to become engulfed in a world of shifting

The author at his desk (1992)

dreams, a flitting fantasy world, a world of multiple illusions, a phantasmagorical world of disconnected consumer-object images;[50] *then to enter the Stockholm Exhibition of 1930 was to be confronted by a single-symbol dream, a dreamworld landscape unified and dominated by the pure and simple line as a line of vision into the future, a monolithic dream-image of a completely planned urban world, an architect-engineered dreamscape heralding a future social world, a concretized dreamworld of highly connected consumer-object images.*

The strict, unadorned, pure and simple line as a line of vision into the future was given one physical manifestation in The Corso, the promenade which began at the exhibition's main entrance and extended forward until it was terminated by The Paradise (*Paradiset*), the largest restaurant building on the extensive premises.

Il Corso. An italian word. Derived from the Latin *cursus*, course, or running race. In Italian synonymous with a city's principal walking street.

Il Corso. An Italian-inspired promenade into the future. A promenade inspired by – among other things – Italian futurism. Futurism, an Italian modernism "marked by violent departures from traditional forms" (Oxford English Dictionary), by a cult-like devotion to the machine and high technology, by the twin convictions that machinery is power and "freedom from historical restraint," that technology is "the solvent of all social ills."[51] Futurism, a modernism whose leading advocates saw themselves as forerunners and partners of Fascism, a modernism that had proved central in the development of those "Rationalist" architects as yet in favor with Mussolini.[52]

Il Corso. A direction-giving street. An avenue of absolute messages. An avenue that went beyond Le Corbusier, Mies and Gropius in projecting a "unified and vital urbanism."[53] A figurative one-way street to a better world. A *senso unico*. Of course! No doubts allowed!

Il Corso. A way leading to heaven on earth. A linear public space for progress-ing toward a new life. A public space where Swedes of all classes and types were to feel at home. "Us" and "them" intermingling. Joined in solidarity. A public space, however, still governed by the realities of the moment. For, patrolling this Eden-bound thoroughfare and the remainder of the exposition grounds were 125 uniformed watchmen and at least 35 policemen.[54] (Stockholm and lesser Swedish cities were still characterized by a pronounced territorialization of class. Off-base, in public spaces, daily-path confrontations could still arouse feelings of uncomfortableness, awkwardness, insecurity or fear.[55] For the bourgeoisie, public surveillance of the "lower" was still deemed a necessity.)

Il Corso. On the whole of one side and part of the other, flanked by a succession of halls and pavilions. Architectural rhetoric of the clean, the pure and simple. First in line, not to be missed, the Hall of Transportation. A part of it jutting out, suggesting the crest of a wave. The wave of the future. And, echoes of Le Corbusier,[56] what else could better serve that function? "Few objects can be said to be so characteristic of our time as the modern means of transportation, the automobile, the motorboat, the airplane ... purely technical products ... little consideration of aesthetic demands in their designing ... no burdensome tradition ... because of technical demands for quickness and lightness every unnecessary decoration must be foregone ... form instead completely dictated by function ... consequently give greater sense of modern design than most articles belonging to the applied arts industry as more narrowly defined ... new designs [now] inspiring within other areas ... where bound by a more or less petrified tradition."[57] And then, the spare slender pillars, glass- and flag-fronted structures, mostly housing the newly inspired. Structures containing not-to-be-ignored visual displays of the rationally designed household good. Structures of illumination for educating consumers of varying economic means for conversion to a new nothing-of-the-bauble, nothing-of-the-gewgaw, no-fooling-around culture of consumption. Structures for optic persuasion, for making desirous the products of some 385 firms. Hall 5: relatively expensive furniture. Hall 8: relatively expensive interior decorations. Hall 9: gramophones, radios and other miscellany. Hall 11: steel-tube chairs and assembly-line furniture. Hall 12: skin and leather goods, musical instruments, office and hotel-room fixtures. Hall 14: gold and silverware. Hall 16: ceramic goods. Hall 17: lamps and lighting fixtures. Hall 19: glassware. Hall 22: wallpaper and linolcum. Hall 26: hand- and machine-made textiles. And so on. Interspersed among these structures, smaller but similarly pure-and-simple-lined mini-pavilions, kiosks and façades broad-casting messages of consumption. Eye-catching colorful walls and unadorned lettering. Each of these buildings devoted to the promotion of a single company, of specific brands of chocolate and mineral water, of specific clothing- and stationary-store chains, of Stockholm's leading daily news-papers. (At some of these firm-sponsored buildings along The Corso – as well as elsewhere – the modernity of the consumer goods in question was underscored by the presentation of immaculate, gleaming-steel, streamlined production machinery in motion.[58])

Il Corso. A commercial concourse where the clean-lined move away from the consumption objects and images of the past is meant to be total. No exceptions! Not even for the hawkers of newspapers or official guidebooks. Not even for the peddlars of balloons, souvenir pins, or lozenges. Not even for the sellers of cigarettes, bananas, oranges, hot dogs, chocolate cookies or baked

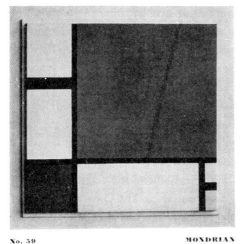

Composition I, oil, 1930

THE STOCKHOLM EXHIBITION OF 1930   119

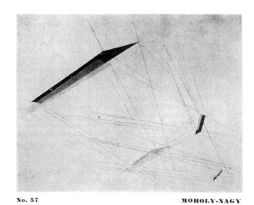

Composition Asc III, watercolor, 1930

Sydsvenska Dagbladet-Snällposten, May 16, 1930

The vertical line . . . is given its most powerful accentuation via the nationally famous "winged" exhibition mast at the Festival Plaza. It points, of course, straight up in the sky. It's not beautiful, . . . but with its naked iron construction it symbolizes the constructive aspects of the whole exhibition.

ANKER KIRKEBY critic for the Danish newspaper *Politiken*, as reproduced in *Dagens Nyheter*, August 1, 1930

No, the Stockholm Exhibition is not built by architects, but by advertising agents. Its decisive character lies not in the fact that nature surrounds and supplements the buildings, but in that intrusive tasteless advertisements suffocate both nature and treetops; among the flower rows, on the house façades, from the opposite side of the inlet – everywhere they scream. . . . Worst of all is the "advertising mast," erected in the middle of the Festival Plaza as the exhibition's sacred token, a metaphor whose steel skeleton is draped from top to bottom with a jumble of advertisements for ties, light bulbs, loudspeakers and chewing gum. It is the pride of the architects and is supposed to symbolize the triumph of modern technology; but one can drolly say that it is the most decadent construction in the entire history of design. In its futuristic-dadaistic confusion, it appears more nonsensical than the most twisted offshoot of rococo or Louis Philippe.

VILHELM PETERSON-BERGER as quoted in *Aftonbladet*, September 5, 1930

The advertising mast . . . seems so trashy and brutal with its messy play of lines and angles and its vulgar advertisements. And remember: yes, of course, this is an exhibition, a big window display, a market, full of market cries that they have tried to tune in a single key. The advertising mast is the psychological key to the whole thing; ugly, violent advertising is {a form of} bragging, and bragging is lower class.

MONSIEUR SERGE "world-famous drawer," interviewed in *Stockholms Dagblad*, June 25, 1930

"It [the exhibition] is matchless. It marks a stage of development which not even the Swedes themselves understand. We must all live in our own era in order to live fully. But the exhibition itself is shaping an era, . . . Modernity has not been as successfully realized, even in France, the homeland of functionalism. . . . Lastly, on the behalf of all French artists, I only want to shout a loud 'Bravo' for one thing at the exhibition."

"Which one?"

"The mast. It is the best thing created in that style, much better than the Eiffel Tower."

"?????"

"Mais oui. I am serious, monsieur. It is the real mirror of our times. And I shout 'Bravo' not only out of politeness, but because it really is so good."

The advertising mast thrust 74 meters into the air. It was the largest advertising device seen in Sweden until that date. By night an illuminating shaft of neon pronouncements. Over 600 lights. Messages of consumption in yellow, red and white. A visual broadcast promoting mass-circulation weekly magazines ("*De 4 Stora*," or "The 4 Big Ones"), radios, lozenges, scarves and chocolates, among other things. Near its bottom – suspended like a gondola – was the official press room, the nerve center of exhibitionary publicity. The exhibition's wings-of-progress trademark was given in supreme version at its crown. At the edge of the lower wing was a rotating spotlight of unprecedented capacity. With a potential reach of 70 kilometers, on clear nights it could be seen in Uppsala.[67]

Dubbed from the outset by folk wits as "*den kolorerade faran*,"[68] most straightforwardly meaning "the colored peril," but also simultaneously hearable as "the illuminated peril," "the sensational press peril," and "the yellow (press) peril."

A token pole of advertisements. A phallus at the center of a space of phallocentric rationality. An erection devoted to the procreation of the new (hu)man. A male penetration of the future, deploying advertisements in order to provide a basic course in modernity; in order to instruct, to foster, to rear, to condition the acceptance of new conditions, to inculcate a no-frills mentality, to eliminate difference, to homogenize; in order to mentally link style and (a new) national identity.[69] An advertising come on, come on (into my home of the future); like most advertising come-ons promising much more than it could deliver. A phallic tower marketing a new idea-logic as well as new goods. A phallic tower possessing none of the anatomical acccuracy of its 1897 exhibition predecessors,[70] but rapier like – long, thin, sharp-edged – with not immediately noticeable extra barbs projecting near its head. A pain inflicting erection disguised as an instrument of pleasure.

A soaring statement. A soaring image inspired, according to several, by Russian constructivism. A soaring structure supposedly inspired by those avant-gardists who created "agitprop art for the Bolshevik cause;"[71] by those post-revolutionary Utopian modernists who sought to bridge the class gap between the artist and the artisan, between the architect and the engineer, and for whom art was a productive agent of social transformation, a symbol for the new, "for a world liberated from injustices, oppression and false pretensions."[72] A soaring symbol apparently more precisely inspired by the likes of the Vesnin brothers' competition entry for the Moscow offices of the newspaper *Leningradskaia Pravda* (1924) and Vladimir Tatlin's famed design and model for the Momument to the Third International (1920), an enormous steel-spiral construction that was to straddle the Moscow River and that strove both to overcome the laws of gravity and to dwarf the skyscrapers of the capitals of capitalism.[73] A soaring steeple seemingly inspired by experimental Russian architects whose buildings "stood for a boundless optimism and a boundless confidence in the possibilities of modern technology,"[74] whose buildings were meant to provide a new space for a world of new social relations. A soaring statement presumably inspired by "Bolshevik" artist-architects and yet a statement pleading for the consumption of goods produced under capitalism. And, moreover, a statement made not too far from "The Humorists's Parlor," a building dominated by large "anti-bolshevik" cartoons that repeatedly irked the Swedish communist press.

A mast to command all, to be seen from below. A mast whose deftly integrated elevator is strictly for sign-repairing purposes. No Eiffel Tower, no breathtaking cityscape views for the many from atop, no provision of an on-top-of-the-world panorama where the city illusorily appears a harmonious totality;[75] but a mast to be dwarfed by, to integrate one with the eyeing masses, to instill a sense of solidarity. A mast which instead obfuscated reality with its overabundance of light, with its multiply blinding advertising messages, with its spotlight suggestion of universal enlightenment by way of consumption.[76]

A mast which was meant to be – and actually functioned as – a center of discussion.[77] A mast to be experienced as "the axle around which the machinery of the entire exhibition rotated,"[78] as the unquestionably most central articulation of this spectacular space of modernity. A mast which served as a metaphor for the centrality of advertising to the entire exhibitionary project; not only the omnipresent advertising of consumption goods and a new ideology of social reform, but the advertising of Stockholm and Sweden to the world at large as well as to the Swedes themselves.[79]

Commencing with the summer of 1928, co-ordinated advertising and publicity campaigns were set in motion so as "to attract as many foreigners as possible"[80] to the exhibition, so as to ensure that products of Swedish modern design thereby gained recognition abroad, so as to make certain that messages regarding "the [Swedish] home"[81] thereby reached an international audience, so as in the process to attract foreign tourists to the natural beauties of Sweden, to the long summer nights and the waterscape charms of Stockholm. As the campaigns increased in intensity, eventually reaching a crescendo in early 1930, over 160,000 brightly colored posters (with their obligatory pure and simple lines)[82] were distributed overseas in sixteen languages, some of them being strategically placed in the dining cars of Europe's most luxurious express trains. No less than 400,000 glossy brochures were distributed abroad in fourteen languages. Short press releases and longer articles regarding the exhibition itself or Sweden's handicraft and home furnishing industries were spread as widely as possible through four operational zones: Norway, Denmark and Finland; Poland and the Baltic states; the English, French, German,

Italian and Spanish-speaking areas of Europe; and the United States and Canada. "Free advertising" in article form was also encouraged through maintaining active contacts with publicists in thirty countries, through offering expense payments and other forms of encouragement both to members of Stockholm's Foreign Press Association and to embassy press attachés based in the Swedish capital, and through inviting selected journalists from Europe, the United States and Japan to view the exhibition while it still was under construction. Local "propaganda" committees established with the help of Swedish diplomatic representatives were used to pave the way for article acceptance by the press and to otherwise disseminate information in large cities throughout Europe. Advertisements such as the following were placed in *The Times* of London and other leading newspapers: "The Stockholm Exhibition is built entirely around the ideal of intensifying the beauty and brightness of everyday life.... Always beauty combined with utility – and modern forms designed entirely for modern needs."[83] Special measures were taken in Europe's major capitals: a neon sign simply reading "The Stockholm Exhibition" was situated so as to be visible throughout London's Piccadilly Circus; a display of exhibition posters and Swedish flags appeared outside all of the major railroad stations in Paris; and a number of neon-lit promotional statements were deployed in Berlin. And, in North America, the exhibition was marketed principally through displays at travel bureaus already acting as agents for the Swedish–America Line of transatlantic passenger ships, through advertisements in publications with an aggregate circulation of nine million; and through articles planted in art and home decorating journals.[84]

If the exhibition was to serve successfully as an advertisement, as a device for selling a pure-and-simple-line version of future Swedish society and a new Swedish identity; if the "broad masses," as well as the well-situated Swede were to be exposed to spectacularly concretized ideology; if a significant percentage of the Swedish population was to be drawn; then it was necessary to advertise and publicize the advertisement itself on a grand scale, to devise a media campaign of unprecedented dimensions. The press release was converted from a one- or few-shot form for immediately impending events to a form of prolonged and incessant informational foreplay, a form for unrelentingly arousing interest in an event to occur many months ahead. With the aid of the domestic wire service (*TT*), press releases and pictorial materials were distributed daily to every newspaper in the country and individuals could now comment: "Never in living memory has the advertising drum been thumped as much as for 'The Stockholm Grand Event of 1930.'"[85] That everyday drumbeat of stimulation was accentuated and amplified by way of press conferences arranged for the Stockholm dailies and major provincial papers, through making information release itself a fanfare event, through staging media events at which Paulsson, Asplund and others either announced new developments or presented building plans and models. In order to further titillate, in order to make promises of satisfaction virtually inescapable, posters were placed in the windows of every outlet of four major national banks, in every passenger car belonging to the state railway and the largest private railroads, in bus terminals as well as rail terminals, in the vast majority of gas stations and automobile showrooms. The wings-of-progress symbol of the exhibition/advertisement became a seeming ubiquity – at least in the country's urban centers of any size – and from July 1, 1929 onward, mail passing through Stockholm's largest post offices was postmarked *Stockholmsutställningen 1930* (The Stockholm Exhibition of 1930). Once the exhibition had actually opened, the seduction of the public was to be completed by posters – again in the required *"Funkis"* style – and newspaper advertisements which spoke of bread and circuses, which announced amusement-park attractions, which trumpeted concert programs, which spotlighted special events. And, for those who might not easily be aroused, stimulated, titillated and finally seduced, there was the club of social pressure, the use of imagery meant to make the reader uncomfortable with the thought of having been left out: "The Stockholm Exhibition of 1930. Over 1,800,000 visitors up to now! Even YOU must see the exhibition!" (Beginning of an advertisement run in seventy-seven provincial newspapers July 7–10, 1930.)[86]

The everywhere visible quality of the advertising mast was symbolic of the fact that other (often innovative) advertisements were everywhere to be seen within the exhibition area. That the advertisement could be an "essential" architectual ingredient rather than a "disfigurement tolerated by necessity,"[87] that the advertisement could be an art form of pure and simple lines, an art form consistent with the exhibition's ideology, was underscored by the presence, near the end of The Corso, of a special "advertising courtyard" – a display which advertised advertising as a contemporary Swedish craft industry.[88] Dozens of firms presented advertisements – all of them approved for stylistic correctness by Asplund and his assistants[89] – either at their own pavilions, shops and "kiosks," or at freely standing installations (including two moored air balloons). Thoroughly modernist statements, spectacular marketing articulations of modernity, were made by modestly scaled concerns as well as industrial giants such as Electrolux, L.M. Ericsson and ASEA (which had not as yet abandoned its swastika trademark). Neon signs, lit by day as well as night, were the favored technique of vision, the eye-catching device of choice. Neon lighting was employed at 135 separate locations, from the Texaco and Shell stations flanking the exhibition parking lot to the majority of shop and kiosk fronts, to the tone-setting mast itself. The high cost of advertising space inclined many leasees to adopt what were then regarded as

daring new devices in order to capture at least a momentary glance if not a prolonged stare. Advertisements were set in motion: spiral-formed signs rotated; lights blinked on and off; product names and messages were electrically unscrolled letter by letter; advertising "shorts" were shown on film screens; a trademark bear nodded in a shop window.[90] Posters were occasionally presented in montage form, as picture puzzles designed to arrest vision, to spur the imagination via kaleidoscope-like juxtaposed fragments, to demand the decoding of their marketing messages, to thereby create a product-identifying sense of achievement among potential customers. Innumerable more conventional, small-scale advertisements occurred either in the form of company-name labels that were affixed to everything from the small rentable rowboats available beside the Festival Plaza to an entire wall of the Park Restaurant, or in printed form on the pages of the two official catalogs, the daily programs and the Official Guidebook. Whether appearing in neon, poster or printed-label form, the typography of the adverisments – with their pure- and spare-lined sanserif and modernized *Antiqua* typefaces – was unmistakably "functionalistic."⁹¹ Moreover, in compliance with the functionalist imperative, any represented human was usually thingified, was reduced to simple naked lines, was made indistinguishable from any represented object, was given a face without distinctive features, was given a machine-like body, was made interchangeable with any other among the masses.⁹²

Straightforward advertising commands echoing the straight-up commanding imagery of the advertising mast:

> Correct taste guarantees high quality.
> Goodrich Tires are best.
> Once tried, always used.
> Never get married without visiting our furniture showroom.
> Eat Swedish cheese.
> The smart woman's underwear.⁹³

The advertising mast, central axis of an advertising vortex, of a whirl and swirl of advertisements that elicited a storm of protest among "a very large percentage of the public visiting the exhibition,"⁹⁴ as well as press critics and even some members of the exhibition's own Executive Committee.⁹⁵ ... Unambiguous vituperative: "highly offensive," "excessive," "distasteful," "jobbery," "hideously ugly."⁹⁶ ... Voices of disappointment: "Advertising dominates all too much at this exhibition."⁹⁷ ... Scornful commentaries: "... it has cost 100,000 crowns to raise the mast. A more idiotic transaction has never been made.... A city should be composed of houses, not of placards, and it should be built by architects, not Barnum."⁹⁸ "I have never seen anything as terrible as this advertising mast.... They seem to have deliberately made ugly advertisements with horribly ugly letters. Nobody with any aesthetic judgment can approve such tastelessness."⁹⁹ ... Most widely offensive, apparently, were those signs interfering with the natural beauties of the exhibition's waterside site. (A notary public claimed he would no longer drink *Pommac*, because the sign for the soft drink obstructed his view across the inlet.)¹⁰⁰ A select group of male establishment figures – professors and business leaders who deemed themselves protectors of nature – voiced fears that the "tasteless" advertisements in question would exert a strong influence on public opinion, that they would undermine efforts to keep the "natural landscape" free of similar publicity, that they would lead to a proliferation of functionalist marketing efforts along country roads, on island beaches, within parts of the "sacred" *Djurgården* area in which the exhibition itself was situated.¹⁰¹ Moreover, the burgeoning of large-scale advertisements already evident along the principal roads leading into Stockholm (as a consequence of traffic generated by the exhibition) would not only spread "uglification" of the landscape elsewhere but, according to the secretary of one of Sweden's national automobile associations, prove distracting and thereby lead to an increase in traffic accidents.[102]

The advertising mast, ultimate statement of the empty promise, of the substanceless mirage, of awakened desire succeeded by dissatisfaction, of the appealing future that really isn't there. The advertising mast as symptomatic (advertising) sign of the times, as seemingly sincere statement that sooner or later proves deceptive to some degree, that – in the extreme case – just may dissolve into flimflam, just may terminate as scam. . . . In 1929, as the eventual opening of the exhibition loomed larger, "a genuine advertising psychosis"[103] took hold in Stockholm business circles. New advertising firms "sprung out of the ground like mushrooms."[104] Many of them were dubious operations whose agents further whipped up the imagination of potential customers, most of them small-firm owners who already had fanciful images of the increased sales that would result from the "anticipated tourist invasion."[105] Unwitting retailers and manufacturers were sold nonexistent space in the "Official Guidebook," biting the bait after being convinced that a competitor would take the same space if the opportunity were turned down. Others were duped into purchasing space on the "Official Map" to be published, on one of the balloons to be suspended above the exhibition or on one of the poster-spots within the exhibition area, even though, in reality, the exhibition's administration monopolized the sale of all such space. Yet others were persuaded to place advertisements in fictitious foreign publications, or in outlets and at locations that could not possibly generate much in the way of additional business.[106] . . . An enterprising hoodwinker advertised that one could purchase rights to "The Exhibition's Grand Gold Medal" for a "registration fee" of ten crowns. No such medal was ever released or even awarded to industrial exhibitors.[107] . . . A con man operating in Alingsås, near Göteborg, managed to secure a down-payment from a victim who thought he was gaining rights to a kiosk at the exhibition.[108] . . . Passing himself off as a valet parker, a deception artist drove a businessman and his family to their expensive hotel from a special long-term parking lot set up for the exhibition and then disappeared with their car instead of returning it to the lot.[109] . . . A smooth talking Stockholmer extracted 300 to 2,000 crowns from numerous individuals who were led to believe they would be given employment as a result of their investing in a firm that was to begin functioning with the start of the exhibition.[110] . . . In the daily press real – but temporary – jobs were advertised by one A. Clintholm on behalf of the "Association for Unorganized Labor." Claiming to offer work at sites "blockaded by labor organizations steered from communist Russia," the group's intent was to build up a reserve of chauffeurs, taxi drivers and car mechanics in order to hold back the wage-increase demands and strike threats that were expected once the exhibition generated heavy traffic increases. A. Clintholm turned out to be a nonexistent person and the "Association," which occupied nothing more than a post-office box, apparently was a front for one or more taxi firms.[111]

Smålands Posten Växjö, May 17, 1930

The most typical machine in the machinery of the exhibition is the enormous main restaurant with its possibilities for simultaneously serving 2,000 seated persons.

Hudiksvallsposten, May 17, 1930

To roughly fashion some clumsy objects that resemble overgrown carrots and call them Adam and Eve appears to be as big a joke upon the Swedish people as cubism was upon painting in its time.

IVAR LO-JOHANSSON (1957), 456

[A]t the main restaurant which was ... like a greenhouse. There sat, ... at one table, a man with two young women whom he puffed himself up for, flattered, and finally paid the bill for. The young women were the same kind of winking dolls as they were in the nineteenth century and they tried to look seductive. They sold sex and competed for the man. – You are strong their glances seemed to say. I am weak, take me. I want to be curly and housewifely and domesticated and have a home where I can lamb in the pen, and then suckle your offspring so they can become cute little lambs. – Any sign of equality between the sexes was totally unnoticeable.

"MONA," in *Stockholms Dagblad*, June 7, 1930

Every day of the week you can tea-dance at The Paradise on a sunlit parquet to the syncopated rhythms of M. Anisz. On an ordinary weekday the public is neither very numerous nor especially elegant, and then you can well go there in dark-blue tailleur. *If you have an elegant silk blouse tucked into a skirt you can perfectly well remove your dinner-jacket. It is really pretty to dance in a blouse, skirt, and hat, particularly at an exhibition tea. In a pleasant way it seems a form of* funkis-*tinged tourist elegance! A beautiful tailor-sewn suit is much more chic than a one-color tea-dress or a flowery rag, with or without arms.*

In the future line of vision extending immediately beyond the mast, beyond the world of desires aroused and promises made, lay the realm achieved upon absorbing the new messages of consumption. THE Paradise. Not the old paradise, once lost. But paradise regained in reinvented form. THE one and only paradise now imaginable. Thoroughly modern. Completely secular. Land of carefree collective delights and unclouded collective satisfaction. The pleasure garden of the new (hu)man, committed to the pure and simple, to

Composition 71, XXX, 1930. Alternative title: Geometrical Intercourse II (*Geometrisk samlag* II)

solidarity and equality. Innocence reattained.

The Paradise Restaurant – with its large glass façade oriented toward the sunset, its prominent canopies and typography, its interior balconies and staircases that combined "grace and refinement with an austerity that might be called classic,"[112] its hints of constructivism and Bauhaus[113] – was a monument dedicated not to solitary pleasure-taking or intimate indulgence, but to enjoyment on a massive scale. The restaurant was reportedly the largest in Europe.[114] Beneath the ceiling-suspended Eden figures by Kurt Jungstedt, beneath the playful 14-meter long snake and the unadorned bodies of Adam and Eve, beneath the cubist-inspired elephant and horse in flight, beneath the air-balloon-like apple, was a dance floor of tremendous dimensions where a thousand or more people could move and sway to the music of a ten-man orchestra.

THE Paradise as architectural manifestation of good intentions: as preview of a socially engineered paradise to come, of a socially democratic paradise to come; as supposedly flawless tableau unmarred by imperfections or serious contradictions.

THE Paradise as garden of comfort: on extremely warm days and cool, starry evenings, the enormous glass surface of the building proved more than illuminating. Either a "hot hell during the day or an ice cold ditto at night" groaned the thoroughly dissatisfied Dane, Anker Kirkeby.[115]

THE Paradise as secular kingdom, as domain free from all that hints of the mystical or metaphysical, as no-religion-here realm of ideological purity: "I hereby wish to inform you that the Executive Committee does not wish 'The Angel Gabriel' to be given shape in the decorative work for the ballroom at the main restaurant" (full text of letter from Gregor Paulsson to Kurt Jungstedt).[116]

THE Paradise as pleasure palace where "a piece of bread, a cup of wine and thou" take on new meaning, where the latter two merge into one, where commodity fetishism is allowed to assume extreme form, where commodified earthly communion is possible: at the restaurant's American bar one could sensuously sip a "Greta Garbo," heroically hoist a "Charles Lindbergh," or, with macho manner, down a "Mussolini."[117]

THE Paradise as democratic haven of blissful consumption, as promised land open to all: by the standards of the time, dining and dancing at the main restaurant was relatively expensive. The cost of a lunch, dinner or even a "bargain"-rate afternoon tea-dance normally exceeded that of a visit to any of the exhibition's cafés, and yet even the prices to be paid at those cheaper alternatives were a matter of widespread public outrage.[118]

Joke: Are you going already?
    Yeah, I'm so damned dry in my throat that I'm taking a taxi into

the city to get myself a drink, but I'll be coming right back.[119]

Much of the restaurant's dining space was booked in advance for sizable groups attending any one of a long series of international congresses, annual meetings and organizational conventions scheduled to coincide with the exhibition.[120] Those who partook in such large-scale affairs, who sat themselves down before an imposing setting that usually included an aquavit shot glass and three wineglasses, were predominantly men in formal attire.[121] ... Those elegantly dressed women and men who danced away the late evening were clearly observable to passing strollers, observable in a space where the romantically private became public, observable in a soft light that turned harsh, in a light that must have evoked envy or anger among those who could afford no more than a look.

THE Paradise as conflict-free retreat, as serene site, as tranquil territory, as peaceful place: reports of discontent among the kitchen staff of 140. The kitchen's "commanding officer," a "foreigner," is accused of using harassment and an assortment of energy- and patience-eroding strategies to precipitate the firing of Swedes, thereby paving the way for the hiring of his fellow countrymen. A strike is threatened, but averted.[122] ... A few days later, a spontaneous strike by more than 100 male waiters, led – according to the conservative press – by "four pure bolsheviks." Disgruntled with a *maître d'hôtel* who apparently was pocketing some of their tips, the men win his reassignment and return to work within twenty-four hours, before their action can be spread either to their female co-workers or to other exhibition restaurants.[123] ... Waiters fired later in the summer claim not to have been refunded the ten crowns deposited when signing their contracts.[124] ... Reworking their various discontents, waiters use a nearby site to offer a 2 a.m. "extravaganza" – "Waiter Willie's Terrible Dream" (*"Kyper Karlsson's hemska dröm"*). They portray themselves as "camels" (beasts of burden), their superiors as (inflated) "balloons," and lampoon Paulsson and Asplund.[125] ... Repeated complaints regarding the name signs and advertisements of shops operating along the ground floor of the restaurant building. The exhibition's vice commissioner is named to negotiate with leasees over size-reduction or removal.[126]

THE Paradise regained in reinvented form as – in the end – paradise lost yet again: one evening, not long before the exhibition ended, a power failure left the restaurant without lighting. The dark was provisionally broken by a few candles, some of them inserted in wine bottles. Suddenly: "*Funkis* seemed unpleasantly banal without all its ingeniously placed lighting effects."[127] ... Towards the end it was revealed that the main restaurant was running at a considerable loss. The high rent charged for the premises was apparently too much to overcome, even though The Paradise had provided "the most distinguished orchestra available;" even though it had become "Stockholm's most fashionable place" during the summer.[128] ... Almost immediately after the exhibition closed, the space of The Paradise was, in effect, temporarily converted to a cut-rate department store. All of the restaurant's cookware and kitchen utensils, all of its china, glasses and cutlery, all of its silver platters, tablecloths and linen napkins were assembled on sawhorse tables and assigned low prices. That which remained unsold after three days was eventually offered on auction.[129] Paradise was relost. Sacrificed to market forces. Sold for a pittance.

Editorial in *Nya Dagligt Allehanda*, May 15, 1930

The educational elements of national and scientific coinage must be counted among the best resources of the now ready exposition.... Thus, one of the features most worth seeing is Svea Rike, *which concretely and instructively illustrates – among other things – our country's tremendous development during the last few decades, not least of all within the economic sphere. The racial biology section included in* Svea Rike *and organized by Prof. Lundborg, will certainly become something of interest {to the public at large}.*

Skaratidningen, May 25, 1930

Without doubt Svea Rike *is the exhibition's most interesting acquaintance.*

Hudikvalls Tidningen, June 26, 1930

It's the best thing in the whole exhibition.

Folkets Dagblad, May 15, 1930

[O]ne must make it absolutely clear to oneself that Svea Rike, *like the remainder of the 1930 Stockholm Exhibition, is entirely an advertisement – and an advertisement for present-day Sweden.*

HERBERT GREVENIUS in *Stockholms Dagblad*, June 4, 1930

Svea Rike's *motley and exciting picture-book makes one big and inspired propaganda brochure.*

TORSTEN KARLING (1930), as quoted in Björklund (1967, vol. 2, 674)

[Svea Rike was] filled almost to the bursting point with [modern] advertising devices.... Without doubt it was just this circumstance which made that separate exhibition so popular. There lights twinkled and gleamed, there {things} whirled and twirled, amusing objects glided, book leaves self-turned, there were articles that the visiting public could play with by hand, free movies, there were long photomontage-flanked walkways, superb advertising texts, more or less easily read diagrams, many of artistic value. There the deepest seriousness was intermixed with the most unbridled jokes of both an

inoffensive and satirical nature. The greatest share of things exhibited there was fun to see and generally had been made readily accessible to an extreme degree.

He who doesn't know his country, doesn't know himself. The fisherman gave birth to the peasant, the peasant gave birth to the priest, the priest gave birth to the burgher, the burgher gave birth to the seaman, the seaman saw the world and his son became a financier, the financier gave birth to the industrialist, and all together they created Sweden, the fisherman his harbor, the peasant his village, the priest his church, the burgher his city, the seaman his ship, the financier his bank, the industrialist his factory, and all of that together is YOU. Travel and find [discover] yourself.

Display caption at *Svea Rike* (PAULSSON, 1937, 171–2)

Svea Rike – There are to be found both romanticism and realism, both poetry and prose. And even the prose becomes ringing poetry when it tells in statistics and diagrams of Ivar Krueger's match trust with its capital of 470 million and 160 factories in 35 different countries, or of Separator's *capital of 124 million and its 110 units throughout the world, or of* Kullager's [Swedish Ball Bearing's] *capital of 130 million and 48 foreign subsidiaries. Such contemporary romanticisms may well have a harsh ring, but it doubtlessly can set the imagination in motion and give the national pride a push.*

Östgöten (Linköping), May 17, 1930

Completely logically, the exhibition's spirit is here, as elsewhere, manifested as a homage to those powers which are presently strong.

Göteborgs Handelstidning, May 20, 1930

And it is fashioned just to be . . . Mem'-ry's fai -rest me-lo-dy!"

Line from "Det är vår sommarmelodi," the "authorized exhibition melody" by Georg Enders and Nils-Georg (1930)

Farthest away along the line of vision extending southward from the advertising mast, something clearly beyond Paradise, something more fabulous than Paradise, was *Svea Rike*, a three-storied building topped by a cylindrical rotunda, a hyperspace where all at once the past was remythologized, the present was glorified and new identities of self and nation were marketed so as to legitimate and hasten the future reshaping of society. *Svea Rike*, a hyperspace where the realm-kingdom-queendom (*rike* in Swedish) of Mother Svea, of the personification of Sweden, of the centuries-old mother of all Swedes, is coincident with the home space of pure and "simple Svea, a young creature, full of bounce, with steel in her eyes, muscles and all of life ahead of her."[130]

The aims of those who conceived and planned *Svea Rike* were straightforward, unambiguous. David Blomberg, an architect and furniture designer,

Bertil Nyström, chief of the National Social Welfare Board (*Socialstyrelsen*), and especially Ludvig "Lubbe" Nordström, a prominent author-journalist, intended to "illustrate how our country has achieved the indisputably high [international] standing that it now occupies within the economic, social and cultural spheres, and what further development possibilities the future might offer."[131] *Svea Rike* was "to strengthen '*l'energie nationale*,'"[132] "to stimulate our sometimes sluggish national imagination and without self-arrogance strengthen our self-confidence as a people and a state."[133] "As far as possible" it was to give a "comprehensive, easily surveyed, concrete and dramatic picture of the realm *we* call ours, that *we* have built up in the world and are firmly determined further to build."[134] It was to speak to "the common people," and especially to the youth of the country, with their highly malleable identities. Its entire contents were – subject to the approval of the exhibition administration[135] – to be in keeping with two historic mottoes included in the pavilion's own catalogue: "Know thyself" and "Don't hide your light under a bushel" (but instead proclaim one's strength and abilities before the world).[136] And, to the extent that it was intended as a project in national-identity reconstruction, as a project in pro-jecting new hegemonic images, *Svea Rike* rested "upon a fundamental paradox – [from modern nationalism], the most international of ideologies . . . [a set of] prototypes, props and models [were to be fetched] for the production of an own national distinctiveness."[137]

Straightforward, unambiguous aims called for straightforward, unambiguous representations, for techniques of vision that would make disjointed data and fragmented "facts" appear totally integrated, that would convey Swedish history as a "totalistic history"[138] which the viewer could easily identify with, easily "insert her own deeds" within.[139] If "dry and boring" "facts and figures" were to be made "enjoyable for the broad public," if they were to convey "an image of our land and people that . . . [was] not only enlightening and authoritative, but also spoke to the [visitor's] feelings and sense of beauty," then the pictorial was to take pronounced precedence over words and numbers, then – "wherever possible" – textual materials, tables and diagrams were to be replaced by hugely enlarged photographs, by pictograms and cartograms, by slides and short films, by functionalist drawings and paintings, by objects in motion, by, in short, "pictures that engrave themselves in the memory."[140] (Distant echoes of Otto Neurath and his proposed museum for "picture statistics," of an insistence that "visual education is superior to word education. *Words divide, pictures unite.*"[141]) All was to appear as a single picture-book, an "amusing and interesting picture-book" with a "pedagogic purpose," a picture-book not unlike the *Biblia Pauperum* of the Middle Ages – "the picture-bible of the poor and untaught,"[142] a picture-book in which hard numbers and cold "facts" were to be transformed into "a visual catechism."[143]

The textual materials employed were largely confined to a spare, epigram-like historical account – authored by Nordström – which began immediately at the entrance ("What have we made of Sweden and what shall we do? Follow the red thread! SEE! PONDER! LEARN!") and extended to the exit. Inscribed on the walls in foot-high letters, and paralleled by a string of red arrows, the text led the visitor from the ground floor to the top floor, from the ice age to the recent triumph of Sweden's multinational corporations and to contemporary Stockholm, the crowning glory of national achievement. As the viewer moved from one historical montage to another, from one panorama of more recent developments to another, and from one thematic display of present circumstances to another, separate tales of linear progress were interwoven into a single narrative of progress.

"*Svea Rike*'s time as a great political power ended 200 years ago, but during the last 100 years its economic power has been slowly built up. Sweden [today] is a leading industrial country.... industry, urban life, international connections demand an intellectually trained and morally mature people [*folk*] who begin to become *Svea Rike*'s principal form of capital. The school becomes the firm foundation of factory Sweden. In the process Sweden's people become one of the most educated in the world, thereby to an even greater degree their own master and nobody's servant.... especially since the end of the World War Sweden has been drawn into what is called A-Europe, or industrial Europe, the nucleus of the world."[144] ... A narrative of progress, a rewriting of history in which traditionally emphasized political events are erased in favor of economic developments, in which national progress and personal pride are welded together, in which the market conquests of Swedish industrial capital are made synonymous "with collective efforts on the part of all Swedes," with an all-inclusive "cooperation,"[145] with the accomplishments of YOU, the viewer, and YOUR ancestors.

Lessons in looking, lessons of consuming interest: cups of coffee and other objects are mechanically moved at one pace to indicate past levels of consumption, at another much brisker pace to reflect current levels.

"People are drawn to the cities, which triple [in population] over 100 years. Stockholm becomes a metropolis of international dimensions, which absorbs the most energetic from throughout the country and leads the way – organizationally, spiritually and materially – in the building up of an ever more globally oriented and worldwide collaborative Sweden."[146] ... A narrative of progress, a rewriting of history in which the rural is demoted and the urban is promoted, in which the bourgeois romantic myth of an uncomplicated, harmonious, friction-free rural past is displaced by a touched up, sanitized image of the urban present as the quintessence of collective improvement and reward-bringing economic growth, as the logical locus for modern functional consumption. A re-educating account whose text and accompanying visual representations were, according to some, throughout pervaded by an undertone of "the World City, Lubbe Nordström's totalistic world city, which has a patent on future happiness, on material happiness."[147]

"Class differences are gradually obliterated, income and wealth are evened out, the general standard of living increases, Sweden's people attain an existence level completely worthy of human beings, a precondition for the highest possible effectiveness and thereby for an ever more distinguished place among nations."[148] ... "In 1870 a general director easily counterweighed the annual income of 20 laborers. Now he counterweighs no more than six, even though his circumstances are in no way worsened. This does not mean that one has taken from the better favored to give away to the worse situated, but quite simply that better conditions and possibilities have actually been created for the latter."[149] ... A narrative of progress, a rewriting of history in which words, numbers and pictures are manipulated so as to make class conflict, political struggle and everyday cultural contestation unproblematic or totally invisible, so as to dramatize the contributions made by the temperance movement and other grassroots "popular movements," so as to cement the interests of individual and society, so as to suggest that what is economically good for the nation as a whole is good for every one, so as to suggest that individual striving, individual industriousness and individual inventiveness benefits all of Sweden.

The travelling time from Sweden to unfamiliar lands by everything from a primitive cart to a modern airplane may be determined by pushing a sequence of buttons. The models of a Viking boat, a more modern sailing vessel and a steamship are juxtaposed with one another, each moving in proportion to its actual obtainable speed. "... giant Swedish steamers built at Swedish shipyards [are] of such high quality that they justifiably gain recognition around the world. But the thoroughgoing transformation that *Svea Rike* has undergone during the present century is perhaps ... [best reflected] by the fact that *the State* has become involved in a growing number of technical activity branches as a consequence of *the realm* becoming increasingly based on steam, gas and electricity. The various activities of the Post Office Administration, the State Power Board, the Air Traffic Department, the Telegraph Service and the National Railways Board all speak their obvious language.... [and] all this newness, the entire technical organization that the State and individuals are providing Sweden, what is its deepest meaning? Freedom!"[150] ... A narrative of progress, a rewriting of history, in which the State as well as corporate capital is depicted as an agent of acceleration, of speeded-up material improvement, of elevation to A-Europe status for each and every one of US, and thereby of one universal emancipation.

An exercise in collective memory (re)construction in which the State as well as corporate capital is situated at the vanguard of technological innovation, is portrayed as a pathbreaker in the application of scientific rationality to the benefit and liberation of ALL. A re-author-ized version of past, present and future in which pride of place is given to reason put in practice, to applied rationality; in which there is no place for the mystical, the unscientific, the irrational; in which religion is almost nonexistent, and all traces of the State Church are eradicated; in which the official Church's centuries-long role as a bureaucratic arm of the State and as an instrument of local social control, its central position in the nation's history, is covered over with silence. A myth of progress from ice-age murk to modern sunlight (by way of machinery and technology, by way of capitalist acumen and collective engagement) in which there is no room allowed for extensive mention of those "dark forces" – religion and the State Church. (This was an idea-logical gesture that was consistent with the Executive Committee's refusal to consider the possibility of religious services at the hall where "ecclesiastical textiles" and other "church art" products were on display.[151]) Amidst the furor that arose over the State Church's absence, *Svea Rike*'s Bertil Nyström claimed the lack of representation was owing to the unavailability of statistics. A church representative countered that there were plenty of statistics, that the question was not one of statistics, but that "*Svea Rike*'s people would have never achieved their place in the sun had they not, under the past nurturing of the State Church, learned self-control, conscientiousness, industriousness and love of humanity.... [that] material culture must have a spiritual foundation and a moral backbone which only can be created by religion."[152]

A side space with a side narrative. A side display on "racial biology," prepared in legitimate, legitimating fashion by Professor Herman Lundborg, chief of the National Institute of Racial Biology in Uppsala. A side discourse in which national identity is buttressed via "scientific" racism. Organized under a rubric lifted from the poetry of Viktor Rydberg: "To Aryan blood, the purest and noble, was I wed by a friendly *norn*."[153] Messages conveyed in brief self-standing bits. "Individuals disappear, but race lives on." "A folk material of good race is a country's greatest wealth." "I am nothing; but my race and my roots and my lineage are everything."[154] A photo- and map-illustrated panel of "racial types" occurring in Sweden, with emphasis placed on "Nordic ('Germanic') Racial Types," with their "tall length, light-colored eyes, blond hair and long and narrow heads."[155] An assemblage of photo-portraits arranged by occupation: farmers here, nurses there; military men here, high-level civil servants there; and so on. A gallery of "culturally prominent figures" – the vast majority of them men – displayed so as to demonstrate that it is not local geographical conditions but genealogy that is most determinative of success, that the "greatest role is obviously played by the better and stronger races."[156] A separate section on the family tree of the so-called "Great Mother of Dalom" (*Stor-Mor i Dalom*), on the descendents of a seventeenth-century woman who married two pastors in her lifetime, on the great numbers of archbishops, bishops, generals, business executives, industrial magnates, mine owners, court dignitaries, leading civil servants, artists, scientists and academics whose origins can be traced back to one Mother of Them All.... A narrative of progress, a rewriting of history, in which women are (Naturally) procreators and men are (Culture) creators, in which every major "racial type" helps make history, but "Nordic" types make more important history than others ("Swedish Walloons," "East Balts," "Lapps"). A genealogy of power which equated accomplishment and societal contribution with heredity, which asserted that Sweden was made of the "right stuff," which wed images of traditional racial romantics and modern social engineers, which was consistent with the functionalist doctrine of the pure and clean, which was congruent with Asplund's functionalist exhibition architecture – white surfaces undefiled by the extraneous. A powerful, power-filled narrative in keeping with the politics of population and hygiene that were welling to the surface in Sweden; with the promotion of "reform eugenics" and legally imposed sterilization designed to weed out the incapable, the undesirable, the unproductive, the perpetually neglectful, the physically and psychically subnormal; with public discourses that insisted on a rational management of the country's health and reproduction.[157]

A side track on a side space. While there are those who might underplay the undertones of racial superiority and Nazi-like racist ideology present in the "racial biology" display, while there are those who might argue that pro-functionalist social engineers were often open anti-racists who came to see population issues in social and technical rather than biological terms, the fact remains that Herman Lundborg had himself written that the mixing of Scandinavian peoples with "less qualified folk elements ... is decidedly abominable," that later in the 1930s he publicly took an antisemitic position and professed Nazism.[158] Whatever the case, the undertones of the display were echoed by less sanitized, less "scientific" expressions of racial superiority and anti-semitism made in connection with other aspects of the Stockholm Exhibition, were echoed by expressions indicative of ideas and idea-logics festering beneath the surface here and there throughout Sweden.[159] ... When a visit to the exhibition was paid by Louis Douglas, the African–American author, composer, singer and performer, one Stockholm newspaper noted that he was "highly cultivated even by western standards" and further observed: "It was one of the black race's, ... and – according to the opinion of many, perhaps all too triumphant – Negro culture's finest representatives who here, for the first time, stood face to face with one of the modern white race's most modern modes of expression in color and form."[160] ... When a picture of a grass-skirted dancer was printed in the country's largest newspaper, the accompanying text read: "This dark girl, who in the white man's eyes is perhaps not so fabulously beautiful, performs every evening in the musical review at the exhibition's amusement park."[161] ... In reporting the anticipated exhibition visit of a number of Japanese, *Social Demokraten* – a supposed voice of solidarity and cosmopolitan tolerance – used the headline: "Yellows attracted."[162] ... Off on the far right, Nazi groups spoke of the entire exhibition as "the Jew Bazaar."[163] They regarded the high food prices charged there as "incompatible with Swedish business customs," as "purely Jewish prices."[164] When some red spots appeared on the Palestinian flag (which flew with its Star of David among numerous other national flags at the exhibition's main entrance), one of their publications commented: "Perhaps it was the flag of another *honorable* nation which had stained the Jew flag so as to mark the *blood* which innocent *white* men have lost during the Jewish race's ruthless expansionist struggles."[165] Their entire attitude toward the exhibition and its "functionalistic soul-killers," toward its glorification of machines and materialism rather than "the blood and soul of humans," was distilled in a cartoon parody of the upper portions of the advertising mast.[166] Atop the mast the wings of progress were placed in front of a Star of David and the wings converted to a barber's razor,

to a jagged-edged murderous weapon.[167] Beneath the razor *"De 4 Stora"* ("the four big one's") were transformed to the *"De 4 Schtora"* (thereby suggesting Swedish with a Yiddish accent), and the names of the four "big" weekly magazines were replaced by the names of four Jews, each of whom was directly or indirectly associated with the exhibition. Josef Sachs was the vice-chairman of the exhibition's Executive Committee, and his supposedly sly and deathly influence on the entire event was underscored by the cartoon caption, "Under the fox-trap," which punned upon the similar sounds of his name and *sax* and simultaneously suggested that the official wings-of-progress symbol was not only a lethal razor, but a fox-trap set for all of Sweden, a trap set by the foxy Sachs. Albert Bonnier was the publisher of *De 4 Stora*, a man synonymous in the National Socialist lexicon with "the Jew-controlled" liberal press. Axel Eliasson was a wholesaling merchant who had obtained monopoly rights for the sale of the exhibition's sole official souvenir, the wings-of-progress symbol in pin form. Eliasson had precipitated a "scandal" by arranging for pin production in Czechoslovakia (note the double-edged word play on the razor: Made in Checkoslovenka), and when the public clamor failed to subside he was forced to award a contract for subsequent delivieries to a Swedish firm.[168]

The final wall-words confronting the *Svea Rike* visitor were a summary of the entire identity-reconstruction project, a distillation of the entire identity-marketing project, a recapitulation of messages that collapsed national and individual progress into one another; that reworked national romanticism and invented a new imagined community of "modern citizens" with a common history and shared future aspirations;[169] that wed individualism and collectivism (and thereby simultaneously served the future agenda of both the Social Democrats and corporate capital). "Don't believe that Sweden is more remarkable than other countries, but don't either believe that it has any less energy, merit or future possibilities![170] Iron, wood and waterpower, money, commercial houses and banks are great things, but still greater is the firm resolution of everybody to stand shoulder by shoulder in order to lift our own country even higher.... [I]f *Svea Rike* is to reach the goal which is the meaning of its entire history – to rise higher and higher in the world's respect – it depends on every man and woman in the country. Fellow countrymen! Brothers! Sisters! The creation of *Svea Rike*'s future depends on YOU!"[171]

On descending from *Svea Rike*'s heights, on returning down to earth via the building's long exterior steps, one was supposed to feel happily proud of one's self as well as one's nation; one was supposed to feel convinced that a good Swede was one who gladly accepted the present and optimistically looked forward to building a bright future by hers or his own hand;[172] one was supposed to feel assured that one lived in a harmonious setting, a setting free of social friction, a setting where all was well and getting better. But, quite

apparently, not everyone did. Far from everyone was predisposed to respond as expected, to read this modern *Biblia Pauperum* according to pre-scription.

A female commentator in a provincial newspaper somewhat ironically observed: "Despite its visual clarity, the much talked about *Svea Rike* made a somewhat confusing impression on a poor woman's brain."[173] And little wonder it did. On the brains of many men as well as many women. And not merely because the cascade of separately clear photographs, maps and diagrams washed away one another. For the world inside *Svea Rike* and the outside world of everyday life did not always jibe very well with one another. Inside one was assaulted with images and (manly) certainty, with unambiguous declarative statements. Outside one daily experienced large doses of uncertainty and ambiguity. Inside the future was unquestionably sunlit. Outside dark clouds were unmistakably gathering on the horizon of the future. Inside was a world devoid of struggle. Outside one did not have to proceed beyond the realm of the exhibition to encounter a world seething with open and latent conflict.

Final lines of a rhyming satirical poem by "Tyst Marie", *Dagens Nyheter*, May 18, 1930

In this temple of statistics,
a picture of Sweden's riches,
of everything of weight and value to the nation.
Here Svea's flourishing figure is shown,
and much one may see – but not everything.
Here the Unemployment Commission is missing.

The future could not have looked so bright to those who were the least bit sensitive to international economic circumstances and their likely ramifications for the Swedish economy. For the economically informed the signs of approaching crisis must have been difficult to ignore, even though Sweden was not particularly hard hit during the summer exhibition by the 1929 Wall Street crash (its home-market-oriented industries had maintained the same high production level of the previous year, thriving domestic business had sustained import levels and export sectors were only beginning to be affected by the collapse of overseas markets). For those already unemployed, the future could not have looked anything but gloomy, even if they did not recognize that many others were soon to share their plight, even if they could not have known that in the months immediately ensuing the exhibition the jobless rate would begin rising sharply (reaching over 23 per cent among trade unionists in early 1933), even if they could not have known that the paramount symbol of Swedish international economic might – the huge financial edifice built by Ivar Kreuger – was an air-castle soon to come tumbling to the ground.[174] Nor could the future have looked so cheery to the politically attuned, to those who could read the writing on the German and Italian walls, even if they did not recognize that *Svea Rike*'s much exalted "A-Europe" had already entered a long slide toward another world war.

The harmonious, let's-all-pull-together world of *Svea Rike* was contradicted by the world of social division and wage-earner versus employer struggles associated with the exhibition itself.... In the months preceding opening day authorities began removing vagrants, beggars and other "undesirable" elements from the streets of Stockholm, placing them in various institutions so as not to offend the sensibilities of anticipated foreign and domestic exhibition visitors.[175] At the same time many Stockholm retailers and service establishment operators were pressuring their employees to take fall or winter vacations so as to make themselves available for the business rush hopefully to be generated by exhibition attendees.[176] Earlier, the collapse of the main restaurant's three-storied iron superstructure resulted in two deaths, seven other serious injuries and labor-versus-capital legal maneuvres based on the claim that the accident occurred because of a forced high tempo of construction.[177] Prior to mid-May there was also an extended confrontation between employers and the taxi drivers who were to serve the exhibition, and the Musician's Union demanded that the Paradise management only hire Swedish dance-band musicians, who were to be paid 20 per cent over the going rate because of the inflated living expenses expected during the exhibition months.[178] ... Once the exhibition was actually under way, it was not only in Paradise that employees expressed dissatisfaction and became embroiled in conflict. There were recurrent general complaints about low wages, lack of toilets, poor eating arrangements and working conditions that often resulted in burns, falls and other accidents.[179] (Of the first 200 accident cases handled by the Red Cross station on the premises, 130 involved exhibition employees.[180]) A brief strike broke out at the café *Gröna Udden* because of the lay-off of union members and the forced payment of uniform and daily admission costs by waitresses and other personnel.[181] The exhibition's private uniformed watchmen threatened to strike because the firm employing them was retaining too high a percentage of the payments made by the exhibition administration.[182] Protesting vigorously, public policemen assigned to duty at the exhibition fruitlessly struggled for extra pay to compensate them for extra food and commuting costs.[183] One month into the exhibition, program hawkers, souvenir pin sellers and other ambulatory vendors who felt themselves exploited began organizing a union local; and in mid-July the fruit sellers who operated on the *Corso*, inside *Svea Rike* and elsewhere went on a one-hour strike over wages and sales commissions.[184] The Musicians' Union went into a furor over the September engagement of Berény's

Hungarian Gypsy Orchestra, threatening to blockade the exhibition if the Executive Committee failed to provide their unemployment chest with 2,000 crowns, and making public charges about Berény's group that dripped with racial superiority.[185] Upset with their pay and with "unnecessary" extra rehearsals during which they were insulted, the Music Review Theater's *Funkis Girls* spontaneously went on strike for one day.[186] Performers in one of the outdoor extravaganzas staged at the Festival Plaza contested the fact that they were not paid for the last minute rain-out on opening night, even though their contracts stipulated compensation for any cancellation occurring after 6 p.m.[187] Receiving little or nothing of the profits being raked in by organizers, sixteen dismayed cartoonists exhibiting at the highly successful Humorists' Parlor withdrew their works, choosing instead to show them at a downtown Stockholm gallery.[188] And so on. . . . The conflict-plagued world of the exhibition not only contradicted the mythical all-is-peace-and-unity imagery of *Svea Rike*. It was also symptomatic of the class skirmishes occurring throughout Sweden. Almost one year to the day after the exhibition opened, in a yet infamous event, four demonstrators and an innocent bystander were shot to death at Ådalen, gunned down by Swedish soldiers who had been posted to protect strike-breakers. . . . Seen then and there, from the future line of vision extending from the advertising mast, through THE paradise, *Svea Rike* (and beyond), from a pure and simple perspective, from a rational(izing) viewpoint: a mere blip of disharmony?

ARTUR LUNDKVIST
(1961), 68

Modernism, that for me was an expression of the spirit of the times itself, those tremors which just then raced along the most sensitive nerves, those visions which fluttered forward in the minds of the imaginative. Purely theoretical contradictions still seemed inessential to me, . . . I thought that functionalism and surrealism could very well complement one another, just as marxism and psychoanalysis.

ELIN WÄGNER
(1930), 20

[T]his new world [of the exhibition] . . . is like a Cubist painting.

----- nuto (a sardonic critic),
Dagens Nyheter, August 20, 1930

[The paintings and sculptures at the exhibition's Park Restaurant] . . . are Funkis architecture converted into civilizing art.

JEAN HÉLION
in the catalog for the Park Restaurant art show
(*International utställning . . .*, 1930) as excerpted in *Stockholms Tidningen*, August 25, 1930

[T]he old "soporific art" is dead, it no longer becomes us to seek an art which is remindful of the fairy-tales that grandmothers tell children who can't go to sleep, or which is comparable to the bugle blast of cavalry troops.

Art is order.	GUNNAR ASPLUND *et al.* (1931), 140
Feeling in life, but not in art. *Romanticism in life, but not in art.*	OTTO G. CARLSUND as quoted in Oscar Reuterswärd (1988), 64
And remember, no symbolism.	OTTO G. CARLSUND as interviewed in *Stockholms Dagblad*, May 9, 1930
Only once in our country has the question of integrating art in society been posed seriously. It happened at the 1930 Stockholm Exhibition with Otto Carlsund's radical exhibit of international post-Cubist art. In a manner that was simultaneously provocative and pedagogic, the art of avant-gardist purism was put forth for comparison with the new architectural form.	PER G. RÅBERG (1980), 12
I thought in my optimism that the exhibit would be a sales success. I took lots of valuable and desirable art works up to Stockholm, all of which were for sale. My agreement with the artists stipulated that their works would be sold and the prices set wouldn't be too high. My commission would be 30 or 33 percent. According to my calculations, first and foremost the major museums would take the opportunity to fill the gaps in their modern art holdings. Furthermore, I counted on the numerous collectors in Stockholm.	OTTO G. CARLSUND as quoted in Oscar Reuterswärd (1988), 94–5
This is not the revolution we need. What we need, presumably, is not so much in the way of new forms as new people.	O---dh (art critic), *Svenska Morgonbladet*, August 19, 1930

Once again. Pure and simple lines. More lines of vision into the future. Another set of pictures to be listened to. Another tale in which consumption and ideological projects are conflated with one another. Another foretelling of Swedish modernity and modernism gone astray. Another (geographical hi)story of the exhibition. This time about pictures at an exhibition. A synedochical saga. About an art exhibit that "began in a frenzy of optimism but ended in tragedy and catastrophe."[189] ... Otto Carlsund. The central figure. Pioneering Swedish modernist. Planner of his own detonation. Bastard child of Swedish industrial modernity? Of Swedish capital abroad? Born 1897 in St. Petersburg. Mother a Swedish governess. Legal father an engineer for Nobel Industries. Biological father (apparently) a Nobel.... Schooled in Norway and Germany before moving to Paris and becoming Fernand Léger's

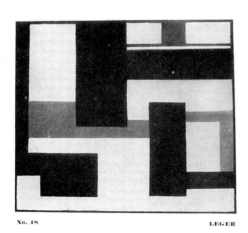

No. 48　　　　　　　　　　LEGER

No. 59　　　　　　　　　　MONDRIAN

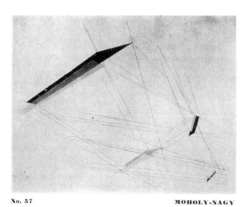

No. 57　　　　　　　　　　MOHOLY-NAGY

No. 14　　　　　　　　　　CARLSUND

"right hand."[190] Propaga(nda)tor of an ideology in sympathetic vibration with that of Paulsson. And thereby with the exhibition as a whole. Under the influence of "constructivism" and its offshoots. There is no border between the aesthetic, the artistic and the artisanal. Art can contribute to the transformation of society. The ideal is the totally true. The objective. The unfrilled. The well functioning.... 1929. Panicked by the wave of surrealism washing over Europe. By the subjective and spontaneous turned loose. By "monkey business" passing itself off as art.[191] He, Theo van Doesburg and three others form *Art Concret*. A journal devoted to warding off surrealism's irrationality and "unrestrained romanticism."[192] To consolidating the position of non-figurative, purely geometrical art. The zeppelin pictured on its first cover an embodiment of ideal form. Of a belief in technology and future progress. The purism of Le Corbusier and Ozenfant pushed further. "Absolute purism."[193] No compromises! ... One Saturday, two weeks before the opening of the exhibition. In the throes of preparing an *Art Concret* exhibit in Paris, Carlsund is summoned by telegram to decorate a long wall at Asplund's Park Restaurant. Monday in Stockholm. Sketch completed on Tuesday. Approved on Wednesday. Painted in short order. "Rapid" – "lines and circles – a rhythm on a wall"[194] representing the surging forward of electrical energy, the propagation of sound, the speed of light.[195] "... so overfunctionalist that it's no painting at all" (Gregor Paulsson).[196] ... An inspiration while painting "Rapid." Why not an *Art Concret* show in the adjacent room? Asplund and Paulsson approached. Convinced of the appropriateness of the glass-vaulted, daylight-drenched space of the "Puck Café" room. Of the entire project.[197] ... Off to Paris to round up paintings and sculptures. Word gets out. Big names and Parisian Scandinavians outside the *Art Concret* group want in. Carlsund decides to expand the show's scope: visitors are to be introduced to "all of modernism"[198] in one room. Sales promises are made to every participant. Upon return, artists elsewhere in Europe are recruited by mail. Using the solicited items as security, financial backing and supplies are obtained.... The show in place. One-hundred and seven "post-Cubist," pur(e)itanical works, including five by Carlsund. Commodities for sale. Visual echoes of the exhibition's architectural spareness. Visual echoes of the exhibition's pure-and-simple-lined advertising art. The only truly international component of the entire exhibition. Giants, near giants and the accomplished but less well-know are represented. Piet Mondrian, Fernand Léger, Hans Arp, Theo van Doesburg, Amedée Ozenfant and Làszlo Moholy-Nagy. Greta Knutson-Tzara, Sophie Täuber-Arp and Francisca Clausen (a fly in the functionalist-like, absolutely pure ideological ointment; she who, while closely involved with Mondrian, defied his ideological stringency and demands for unquestioning obedience, refused his straight and narrow imprisoning bars, introduced other images and hints of the surreal).[199] ... Drunk with optimism, with his ability to remake Swedish tastes, Carlsund tosses borrowed money about. He publishes a glossy catalog. And prints invitation cards, elegant stationery. And caters a preview showing for museum officials, art critics, participating artists. And a banquet afterwards.... And then disaster. More or less complete calamity. More or less total rejection. A cascade of laughter and abusive words. "Most laugh at the seven isms [Constructivism, Purism, post-Cubism, etc.] represented at the Park Restaurant exhibition, some get angry about them, few understand them and even fewer wax enthusiastic about them.... There is something decadent about this art."[200] "... most of it is confusingly anxious, mechanical and impersonal."[201] "I have extreme difficulty in understanding [the claims of Le Corbusier and Ozenfant that paintings and sculptures like these can] ... suit ordinary people – to the extent they are not feelingless robots."²⁰² "Is it as a [market] fair amusement that the exhibition has arranged for us to see this new salvation? In that case perhaps it would be more at home in the rather dreary Amusement Park area. There its unimaginativeness would be right in style."²⁰³ "A gentleman from Tranås wandered about, wondering if he was in the Humorists' Parlor."²⁰⁴ Cartoon caption (a well dressed couple standing before a parody of one of the Park Restaurant paintings): "This is called *Art Concret* – a quadrangular canvas, painted with circles, straight lines and triangles – the producer is a genius and the buyer is a meathead."²⁰⁵ ... No "meatheads" available. Not a painting sold. Virtually all of the catalogs unsold – ultimately destined for the garbage dump. Carlsund suffers a nervous breakdown a few days before the exhibition closing. Many of the art works appropriated by Carlsund's financial backers and suppliers. Others given away for a song at a sale improvised in order to meet costs. Carlsund, unable to return either money or works to any of his artist friends, refuses to answer or even read the stream of correspondence from them. A silent *adieu* to his collaborators. Followed by a practical *adieu* to *Art Concret*. To that "solitary and frozen swallow which first, only much later, presaged a new summer."²⁰⁶ ... After words bemoaned afterwards. "My luckless, unhappy exhibit."²⁰⁷ "... [E]verything went to hell." "I didn't succeed in getting as much as one percent of the visitors to adjust their eyes and see that it was an art exhibit I [had] arranged."²⁰⁸ ... And why was it that people didn't see? Didn't choose to hear other voices with their eyes? Didn't buy? ... How much of the failure can be attributed to "insufficient advertising"? How much to the "exhibit's exclusiveness," its appeal to no more than a thin layer of "the art-interested [public]"? How many of those with a willing eye or wallet were were put off by the "primitive hanging"? By "the unsuitability of the café locale"?²⁰⁹ By the compressed arrangement of the paintings? By the lack of breathing space for any one of them? By the difficulty of viewing items stacked three or four atop one another? By not being able carefully to scrutinize works substantially above eye level? By the tables, chairs and chattering café guests that prevented close examination, that discouraged thoughtful contemplation? Was the entire arrangement so jumbled that many could make no distinction from another Park Restaurant space – "The Label," a room whose walls were blanketed with commercial labels? Or was it mostly a matter of a paradox at work? Nonfigurative artistic efforts that were "too new, too incomprehensible for the Swedish bourgeoisie,"²¹⁰ and yet "old hat" from a continental perspective, that were "dying echoes from the years of struggle when modernism emerged"?²¹¹ Were Swedes not yet ready for what had been avant-garde five or more years before? Was it simply the case that "Swedish ignorance of what was happening out in Europe had long been total"? And was not about to be disturbed? After all, was it not so that the tone-setting Stockholm daily *Svenska Dagbladet* had used put-downs when it first presented the *Bauhaus* after waiting until "it had been at work seven years, until three years had passed since the famed Bauhaus Exhibit of 1924, [until] nine years after the founding of *de Stijl* in Holland"?[212] Or is it possible that a passive reception for avant-garde art in Stockholm was long before doomed by the absence of cabarets? By the pre-World War I absence of a "cabaret culture," of breeding-ground and foothold sites such as those found in Paris, Barcelona, Berlin and other European metropolises?[213]

Functionalism means that a chair is a machine to sit on.

Popular definition, parodying Le Corbusier's "a house is a machine to live in," as quoted in *Skånska Social Demokraten*, June 14, 1930

That which immediately struck me as characteristic of the great majority of the visitors was their conservatism, nostalgia, romanticism. When they saw everything in the light of the new era, their eyes became round and screened out like those of owls. They seemed unable to tolerate the clarity.... They became disoriented as soon as they entered the flat-roofed, chimneyless functionalist houses – which seemed to abolish the cold and snow of winter – and tried to use their eyes to find their way to the shiny steel-tube furniture. "Doesn't that look nice?" they said of an old rocking chair with nauseating pillows that had been put in the corner as a bad example. They couldn't stand the steel beds, but still had the old sofa-beds and shaggy couch-beds left in their mouldy minds. "Who wants to sleep in that?" they said. And when they came to the glass police station it was the same thing. "Who would want to lie in a clink like that? Give me the days of my youth."

IVAR LO-JOHANSSON (1979), 455

The unreality of the exhibition area – a city where nobody lived at night but only during the sunlit day, where the houses and objects were only for viewing, but not for using – suddenly filled me with fright. The great steel birds seemingly tried to take wing, but they couldn't. Visitors to the modern seemingly wished to breathe fresh air into their lungs, but they only got coughing attacks. A thought occurred to me: humans invent new things faster than they can learn to use them. Right below the glistening steel constructions common house flowers grew in the garden-beds, ... [flowers] that had to be watered with ordinary country water in order to prevent their dying. No

IVAR LO-JOHANSSON (1957), 6

matter how hard I searched, I couldn't find the new [hu]man anywhere.

If nobody was buying the art Carlsund was exhibiting at the Park Restaurant, who was buying anything else offered at the exhibition? Once again: who was buying into a program that served the interests of industrial capital and yet was congruent with the Social Democrat's imagery of a "People's Home"? How many among the attending millions were buying any or all of the exhibition's high-flying rationalities and offerings of modernity? How many were coming only for bread and circuses? How many were coming after 7 p.m., when the cost of admission was cut in half, when the amusement park was in full operation, but when the exhibition halls and pavilions were closed? How many among the throngs were identifying with the new housing, the new household furnishings, the new prescriptions for urban design and urban living? How many among those gender-, class-, occupationally- and regionally-difference-iated millions were absorbing the messages offered, were allowing the meanings offered to become central to their identities?

Or: how many were buying the spectacle of pure and simple lines? How many were buying either the goods or the future line of vision offered by the *Corso*? How many were buying the future line of vision extending from the advertising mast, through THE Paradise and *Svea Rike*, and onwards into the historical beyond? Who was actually buying into this introductory course, into this new would-be hegemonic discourse, into this new cultural politics, into this new alignment of power and aesthetics, taste and style? Who was succumbing, was being partially reconstituted as a subject, was buying into a new grammar of modernity, a new set of "general ideas about how, when, where and why to be modern," a new set of "ideas about progress and enlightenment" coupled with a new "iconography and symbolism"?[214] Who was being made over into a "modern" and "rational" consumer? Who was experiencing the world of the exhibition as Paulsson and his associates desired, was being educated as desired, was reidentifying herself through the acceptance of new meanings, through the formation of new desires, through buying the congruence of new forms of advertising, architecture and art? Who was buying into this future world where consumption was to be a form of social engineering, this future world of the shiny, the streamlined, the "high-quality," the efficient, the practical, the functional? Who was buying into this brave new world by discarding, repressing, rejuxtaposing or reworking meanings already embedded in their ways of life; by adopting new forms of consumption which demanded a rearrangement of everyday life – a rerouting of daily paths, a reroutinization of daily time-spaces;[215] by accepting clearly defined differences; by beginning to make new distinctions?

Who was buying the uncompromising pure and simple lines, the future lines of vision? Who was willing to part with their money and their already held meanings? Not those who turned their backs on the goods displayed in the halls along the *Corso*; who by August caused the Executive Committee's vice-chairman to complain that "exhibitors almost without exception have had little success in selling exhibited objects."[216] Not those who thought that steel-tube chairs and other metal furniture would be too hot in the summer, too cold in the winter, "unbearable in front of a fire."[217] Not those who could be amused by a cartoon showing "the inventor of steel furniture" at work while seated in a well-cushioned traditional armchair, in a functionalist symbol of "backward" living, in a chair not well suited for modern times, for sitting alertly, for quickly springing to life.[218] Not those who could identify with the message of another cartoon showing five fetishized chairs, five highly ornate chairs with faces laughing themselves to tears over a steel-tube chair whose face is in tears of misery over its reception.[219] Not those who refused to purchase the functionalist lighting fixtures and furniture moved directly from the exhibition to the up-scale NK department store; who instead chose the more traditional items offered by that costly retailing outlet.[220] Not those who abandoned the gramophone section of Hall 9 rather than swelter in afternoon temperatures around 38°C (100.4°F).[221] Not those who found the young women relentlessly attempting to sell the wings-of-progress souvenir pin nothing less than "absolutely insufferable," who likened their "attacks" to those of "prehistoric harpies."[222] Not, very likely, many of those among the great majority who would have nothing of the main catalog, the official guidebook, the catalog for the housing exhibit or the daily program. (Cold statistics: over 4.1 million attendees but only 4,510 housing catalogs sold, only 5,767 main catalogs sold, only 63,881 official guidebooks sold, only 70,748 daily programs sold.[223]) Not, in all probability, many of those who found their exhibition visit a jolting experience, who literally were blind to the material manifestations of the exhibition's ideology, who could not see the light, who injured themselves by walking directly into glass doors or walls.[224]

Who was buying the uncompromising pure and simple lines, the future lines of vision? After visiting the model homes and apartment interiors of the housing exhibit near the amusement park, who was willing to consume the new alternatives, the new "solutions" for the housing problem, the new visions of housing as a form of consumption? Who was buying the strict functional specialization of space called for in the new housing units? Who was buying the social engineering of family life, the norms for family social interaction being put forth by functionalist architects, by self-design-ated experts? Who was buying the notion of a large, amply windowed living room as *the* family gathering place, as *the* center of the family universe, surrounded by satellites of secondary importance: a laboratory-like (supposedly) time-saving kitchenette, or efficiency kitchen; separate bedrooms, each with space for little more than the beds they enclosed; and a minute bathroom, with washstand, toilet and tub or shower directly abutting upon one another? Not those who had been schooled to regard homemaking as a moral imperative rather than a modern and scientific activity, who could not understand why industrial science and management belonged in the kitchen, or why or how what later came to be termed "Taylorism for housewives" was to work.[225] Not those who could not understand either why – hot stoves or not – children shouldn't play in the kitchen under the watchful eye of a cooking mother, or why the traditional kitchen sofa shouldn't be available for napping purposes.[226] Not Ingeborg Waern-Bugge (Sweden's first female licensed architect) who regarded the kitchens designed by her male colleagues as "unrealistically small."[227] Not any of the other women who bristled at male-monopolized "solutions," who did not wish to be instructed by those supposedly above them, who were suspicious of anything that smelled of pater-nalism. Not the more radical women who found much male-dominated functionalism too conservative, who believed that employed working-class women would be better served by more extreme forms of functionalism, by apartment houses with collective facilities, with facilties that facilitated a more equitable sexual division of labor.[228] Not those on the left who were able to see that functionalist apartment designs were not revolutionary; that such "solutions" were nothing more than a technically oriented, reformist palliative for deep-seated social problems; that such "solutions" addressed the efficient, rational use of small spaces rather than the redressing of large social inequalities.[229] Not those working-class women and men who remained strongly attached to their own practiced spaces of intimacy, who believed in the preservation of a "fine room" (*finrum*), or parlor, a seldom used space "with its plants, mantlepiece clock, and lace-decorated sofa;" in the preservation of a "silent and well-kept" space removed "from the drudgeries of everyday life;" in the preservation of a meaning-charged, special-occasion-only space that "radiated dignity" even if conditions were so crowded that there was nowhere else to sleep but in the kitchen.[230] Not those who, when occupying a *"Funkis"* apartment a few years later, insisted upon converting the living room into a seldom used "fine room," in cramping their activities into the kitchenette, in darkening the big windows with thick curtains and large-leafed plants.[231] Not those who believed that the provision of in-house plumbing for working-class families was not enough, who believed that a bathroom ought to be something more than a crowded cubicle.[232] Not those among the better situated who thought the more luxurious model homes "interesting . . . but not something I would want to live in."[233]

Who was buying the uncompromising pure and simple lines, the future lines of vision? Who was willing to pro-ject the exhibition's functionalist project into the future, to buy its messages of solidarity and jointly marching into the future, its messages of the new "(hu)man" and the "new society"? Not those who clung to their romanticized images of the past and past traditions, their old historical myths of Gustav Vasa and other heroic kings, their myths of a once friction-free rural existence. Not those who clung to their old identities, to their bourgeois stance on the totally independent individual. Not those who refused to stem the (creative) destruction of the exhibition or any part thereof. Not those wealthy persons and institutions who got cold feet about permanently preserving the contents of *Svea Rike*, who grew reluctant about providing the funds necessary for shifting them to a museum or some other location, who lost their enthusiasm for expanding the exhibit for presentation at the 1933 Chicago World's Fair.[234] Not those who would not rally to calls for converting THE Paradise into a giant studio for young painters or for leaving the immensely popular planetarium and aquarium physically intact.[235]

Who was buying the uncompromising pure and simple lines, the future lines of vision? Who was buying and thereby demonstrating their capacity for enlightened citizenship? Not those who, right from the beginning, symbolically reworked their discontent with the entire business by renaming the exhibition area, not merely as "*Funkis*," but as "Box City" [*Lådstan*], "The Aspen Grove" (*Asp-lunden*, a play on architect Asplund's name), or as

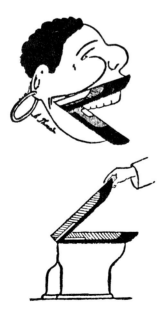

"The Mast-Sting" (or "-Bite," "-Stab," or "-Blow" [*Mast-hugget*, at once a comment on the advertising mast and a play on a district in the city of Göteborg]). Not those who more cuttingly labeled the exhibition as "The Barber Shop" (*Rakstugan*, a triply pun-ishing term, referring to the exhibition's straight-lined functionalist architecture [a literal translation of the word's two roots would be "The Straight Cottage"], to the perception that the public gets cleanly shaven, and to the resemblance of the exhibition's wings-of-progress symbol to a barber's razor, or *rakkniv*).[236] Not those who culturally contested the insistent clean-linedness of the exhibition's architecture and consumer goods, who spoke of the area's buildings in general as "the packing crates" (*packlårna*) or "the steamboat decks" (*ångbåtsdäcken*), who verbally classified the single-family homes of the housing exhibit as "the madhouses" (*dårhusen*).[237] Not those who constructed their own meanings of the exhibition's physical and ideological constructions, who saw through the objectification of the subject and the subjectification of the object inherent in the exhibition's conjoining of consumption and social engineering, who knew that one of the things they were seeing was thingification, who had their own counter to commodity fetishism, who knew what was meant when a cartoon caricatured Paulsson as the (Exhibition) Commission's office building (situated at the head of the *Corso*), Asplund as the Planetarium and the publicity-chief Brilioth as the international press facility.[238] Not those who experienced the exhibition as a shock of (ultra)modernity; who were shocked by what was depicted as the "ultramodern," as "the everyday life of tomorrow;"[239] who experienced a threat of disruption and discontinuity, who consequently struggled over the meaning of "progress." Not those who gave voice to their symbolic creativity[240] by referring to the official wings-of-progress symbol, to the official vision of the future, either as "*Sachsen*" (wittingly or unwittingly expressing anti-semitism and notions of a Jewish plot),[241] "the pocketknife" (*fällkniven*) (suggesting something dangerous and yet of little significance), or the "toilet lid" (suggesting something full of shit, something that ought to be flushed away).[242] Not those who took mirth at, or were verbally inspired by, cartoons re-presenting the wings-of-progress symbol as a toilet lid, the lips of a "primitive" black woman – perhaps Josephine Baker – (here the lid of racism once again lifted), a beggar's arm, a ski-jumper's skis, and so on.[243]

In not immediately buying the pure and simple lines, the future visions; in not bringing the 1930 Stockholm Exhibition to the same end as its 1897 predecessor; in not fulfilling the fears of the authorities and the press, not ending the exhibition with a riot in which nearly every thing, every good, within reach was plundered;[244] in refusing to take home a souvenir, to either purchase or poach an object-ive confirmation of collective memory, the vast majority of attendees in effect took a physical and symbolical distance, in effect physically and symbolically rejected messages that conflated new consumption forms and national progress, in effect chose absence over presence, in effect opted for inaction and silence as a mode of signification, refusing to allow new goods and meanings to become a part of them, to become central to their individual and collective identities.

Of course, not everybody refused the pure and simple lines, the future visions. Some were willing to buy into the messages being put forth by Paulsson, Asplund and the cosmopolitan functionalists around them, by "a handful of men, almost all of the same generation – around 40 in 1930."[245] Some buyers were to be found among those who were being absorbed in the newly emerging bourgeoisie – among "teachers, white collar workers of different levels and businessmen,"[246] as well as among certain intellectuals, certain members of the labor movement's upper-leadership echelons. But – seeing things and hearing messages in different ways out of habit(us) – the vast majority were not: most women and men born into Oscarian (Victorian) middle-class values were not; most farmers and their families were not; most farm laborers were not; most members of the urban working classes were not.[247]

As yet.

EPILOGUES

I am surely not a circus director.

———

The exhibition as source of contamination? As infection introducer? As bearer of consequences not apparent at the moment of contact? ... By the time the exhibition had run more than half its course the "traffic" expansion in Stockholm resulting from it had apparently generated an increase of 25 per cent or more in the occurrence of gonorrhea and other venereal diseases. City authorities chose to deny there was any substance to the evidence,[248] while the establishment press chose to remain silent on the entire matter.

The exhibition as generator of eventual transformation? As promotor of acceptance of "the [further] invasion of capitalism's instrumentality into the fabric of everyday life"?[249] As ultimate propagator of new individual/state, person/commodity, subject/object relations? As delayed-reaction producer of the "new man," the "new woman"?

GREGOR PAULSSON
(*Nya Dagligt Allehanda*, September 29, 1930), in emphasizing that the exhibition would not be prolonged one weekend beyond its originally scheduled closing date

Architectural composition or *The Factory*, a portrait of Margareta Alexandra Frölich, by Otto Carlsund, drawn in 1931, not long after recovering from his breakdown.[250]

"It's too bad that all this which has cost so much labor and money will be dismantled in the fall ..." said somebody during a walk through [the yet uncompleted] exhibition area.

"On the contrary," came the immediate response, "when the Eternit sheets are plucked down, then the energy that has been concentrated here will be set free in the universe and do its work. What more can one ask?"
It seemed as if it was already free in the universe.

ELIN WÄGNER
(1930), 23

As modernity was [further] materialized in objects, hopelessly outdated things became symbols of shame.

ORVAR LÖFGREN
(1991b), 11–12

From [the] 1930s onward, people in Sweden would become the most disciplined and well trained of all the domestic animals.

IVAR LO-JOHANSSON
(1974) as quoted in Jonas Frykman (1981), 61

Eventually resistance broke down among the great majority. Eventually the reluctance of most came to nil. Eventually the symbolic discontent gave way to new meanings. Eventually identities were reshaped. Eventually functionalism triumphed. Eventually functionalism assumed a hegemonic status. Eventually the interests of both the Social Democrats and large-scale Swedish capital were served. Eventually there was no shortage of buyers. For the pure and simple lines as well as the future visions. For the commodities – even the nonrepresentational art works – as well as the messages. For the cleanly engineered goods as well as the social engineering – the shift of social responsibility from "individual" to "society," from (what were design-ated as) identical women and men to the State. For the housing as well as the regulated con-form-ism and the ideology of solidarity. For modernity as well as technocratic re-form-ism.

Eventually *"Funkis"* became some-things natural, self-evident, taken for granted. Eventually the pure and simple lines of functionalism, its array of meanings, became as much a matter-of-fact presence on the architectural landscape of Sweden's cities and suburbs as on the commodity-scapes of the country's domestic interiors, as on the "invisible maps" characterizing the internal worlds of Swedish women and men.[251] Eventually people were reoriented, their habit(u)s restructured, by the severe economic and social crisis that immediately followed the exhibition.[252] Or, they learned the lessons of "rational" consumption that were taught and retaught in the local-level "study circles" sponsored by various Social Democratic institutions. Or, following the wartime austerity of the 1940s, when pent-up demand and expanded purchasing power were jointly released, they not infrequently bought into the new cornucopia of functionalist-style goods – in much the same way they bought into the torrent of Social Democratic reforms, the technocratic plans and new techniques of power that intruded into everyday life – because of shame, because of a fear of being different (because of the most traditional form of Swedish social control assuming a new guise), because of a fear of being *un*modern, of being left behind.[253]

But all of this involves another set of (geographical hi)stories. (Geographical hi)stories not to be told here. (Geographical hi)stories whose nonclosured closing episodes resonate with the (geographical hi)stories told here. (Geographical hi)stories whose nonclosured closing episodes, whose ongoing sequels, demand to be read intercontextually, to be read through the spectacular spaces and articulations of modernity associated with the Stockholm Exhibition of 1930.

How is one to reread, re-view, re-hear, re-reread
the (geographical hi)stories of the Stockholm Exhibition of 1930?

How is one to do so
as the future lines of vision made visible and given voice at that exposition
    are rerejected,
as the pure and simple lines are no longer bought,
as the meanings associated with those lines lose their moorings,
as those lines disintegrate into pointlessness,
as a protracted era which symbolically began with those lines
    now undergoes dramatic decomposition and confusingly unravels,
as solid(arity) success melts into the thin air of disillusionment,
as the welfare state writhes in crisis,
as new hollow promises of a consumption Paradise
    are repeatedly recited by "new-liberal" prophets of the "free market"
                      and the Common Market,
as Swedes are confronted by a disjuncture
    between once central meanings and current meanings,
as they culturally rework once rejected but later triumphant meanings,
as they suffer the dis-ease of individual and collective identity crises?

What is to be reread from the conflation of a new aesthetics of consumption, "progress" and the creation – the social engineering – of a new Swedish (hu)man?

How are the exhibition's advertisements, its messages, its pronouncements, to be re-viewed?

How are the manifestations of rank commercialism, class tensions and racism to be re-heard?

How is one to reread the attempts to reconstitute collective memory, to reinvent and remythologize the past, to simplify and manipulate history?

How is one to re-view the unwavering commitment to the rational and the technical, the insistence on the regulation and control of every exhibition detail?

How is one to re-hear the initial refusals to buy?

How are all of the (geographical hi)stories of the 1930 Exhibition to be re-reread, re-reread, re-reheard, re-reheard, in the context of their manifold resonances with contemporary, here and now, Sweden?

Today's [Swedish] society has its determinations in the past, but the past which made the present possible could also have had other consequences.... [T]here is no arrow-straight connection between, on the one hand, historical contexts, forces and forms and, on the other hand, the occurrence of concrete events.

GÖRAN THERBORN
(1988), 47

The road from 1930 to 1984 and beyond was paved with good intentions.
Where else could that pure-and-simple lined, that undeviating, unswerving road have led?
Were there no alternative routes?
Could the modernist masculine rationality which laid it out find no other opening?
Could a pure-and-simple-lined "People's Home" have been constructed,
could equitable care and freedom from want have been provided,
    without occasional bureaucratic intrusions upon freedom,
    without grim paradoxes and contradictions,
    without social engineering that objectified the subject
        and sought to suffocate difference and spontaneity,
    without colonizing the spaces of everyday life with new power relations,
    without a dense grid of control,
    without the intensification of pure-and-simple-lined authority?

LEIF NYLÉN (1991)

Le Corbusier never came to the Stockholm Exhibition, even though he was regarded by many as its guardian angel or evil genius. Le Corbusier was invited but by mistake was given a second-class sleeping car ticket – sullen, he turned back at the Gare du Nord.

FIGURE ACKNOWLEDGMENTS

Photograph of couple descending steps from *Svea Rike* by courtesy of *Arkitekturmuseet* (The Swedish Museum of Architecture). All other exhibition photographs by courtesy of *Stockholms Stadsmuseum* (The Stockholm City Museum). Leger, Mondrian, Moholy-Nagy and Carlsund paintings from the catalog of the "International Exhibit of Post-Cubistic Art at the 1930 Stockholm Exhibition" by courtesy of *Norrköpings Konstmuseum* (The Norrköping Art Museum). All other figures from published sources indicated in the text. Carlsund's *Architectonic Composition* by courtesy of Oscar Reutersvärd.

NOTES

1 Buck-Morss also lists expositions in Paris (1931), Chicago (1933), Brussels (1935), Paris (1937), Glasgow (1938) and San Francisco (1939).
2 Cf. Therborn (1988).
3 Löfgren (1991a), 105.
4 Cf. Ambjörnsson (1991b) and Andersson (1991). The collective identity crisis experienced by many Swedes also stems from the gap between their self-image as international altruists and their complexly mixed feelings about Third World refugees and immigrants, about the surfacing of racism in their own country. For further details see pp. 219–21.
5 The importance of the 1930 Stockholm Exhibition to an understanding of contemporary Sweden has in various ways been suggested by Frykman and Löfgren (1985), Hirdman (1990), Nilsson (1991) and Nylén (1991), among others.
6 The Swedish Handicrafts Association had been responsible for the Stockholm Industrial Arts Exhibition of 1909, an influential Home Exhibition at Stockholm's Liljevalchs Museum in 1917, a major section of the 1923 Göteborg Exhibition and a number of displays at international expositions held in Switzerland, Italy and elsewhere (see Råberg, 1972, 20–1). From its nineteenth-century outset the association had sought further collaboration between artists and manufacturers so as to counteract the quality decline and ugliness supposedly resulting from the mass production of home furnishings and household utensils. A summary account of the association's role in the origins

of the exhibition is contained in Paulsson (1937).
7 Paulsson (1970), 106–12. Sweden won more gold medals at the Paris Exposition than any other country, save France (Kommittén för utredning av allmän svensk utställning..., [hereafter Kommittén] 1928a, 5).
8 This exhibit, which later moved on to Chicago and Detroit, helped give birth to the often abused American expression "Swedish modern" (Björklund, 1967, vol. 2, 679).
9 In order to meet construction costs and other budgetary demands, the then considerable sum of 4.6 million crowns had to be obtained (Rudberg, 1985, 99; *Svenska Dagbladet*, January 31, 1928 [this and all subsequent newspaper and magazine references obtained from the clipping collection that forms part of the extensive 1930 exhibition archive held at *Stockholms stadsarkiv*, hereafter SU-1930]). The exploratory committee included Gunnar Asplund, the exhibition's eventual architect, as well as Paulsson.
10 Kommittén (1928a), 4.
11 Kommittén (1928a), 5–6.
12 Kommittén (1928a), 6. Also see *Svenska Dagbladet*, February 1, 1928; *Göteborgs Handelstidning*, February 8, 1928.
13 *Stockholms Dagblad*, July 28, 1929.
14 Kommittén (1928a), 7.
15 *Dagens Nyheter*, January 31, 1928; *Svenska Dagbladet*, January 31, 1928.
16 Kommittén (1928b).
17 Cf. Rydell (1984), especially 235.
18 *Program och bestämmelser* (1928), 12–15; *Officiel huvudkatalog* (1930), 13–16.
19 See Chapter 1, above.
20 This catch-all division subsumed items such as street lampposts, telephone and electricity poles, park benches, drinking fountains, and boats, cars, buses and airplanes as well as their interior decorations (*Program och bestämmelser*, 1928, 7–8).
21 The monopoly given to Frans Svanström & Co. for the production and sale of postcards as well as picture albums "within the exhibition area and on the open market" soon resulted in a legal tangle with a would-be competitor. *Nya Dagligt Allehanda*, June 15, 1930; *Svenska Dagbladet*, July 10, 1930.
22 *Ordningsföreskrifter och andra upplysningar för utställare och hyresgäster.*
23 *Program och bestämmelser* (1928), 1–25; *Social Demokraten*, July 18, 1928. The Paris Exposition of 1925 had similarly barred "all copies or counterfeits of historical styles," had similarly banned all that did not show "clearly modern tendencies" (Greenhalgh, 1988, 164).
24 *Svenska Dagbladet*, February 1, 1928; Kommittén (1928a), 6.
25 Undated letter, Gregor Paulsson to all exhibiting firms (SU-1930, miscellania, box 44).
26 Paulsson (1937), 1, 4.
27 Kommittén (1928a), 4, 8; *Program och bestämmelser* (1928), 1.
28 *Svenska Dagbladet*, May 29, 1928; Kommittén (1928a), 6.

29 Kommittén (1928a), 7.
30 Råberg (1972), 146; Paulsson (1928), 9. The standard literature on the Stockholm exhibition of 1930 generally ignores its economic objectives. That Paulsson and those in command of the exhibition were more than a little preoccupied with increasing the sales volume and profits of Swedish manufacturers is repeatedly borne witness to by the minutes of the Executive Committee's meetings. For example, in a memo attached to one set of the minutes Paulsson contended that the Hall of Transportation might well bring Sweden's car and bus manufacturers a higher level of output and an increased ability to compete with other producers (*Verkställandeutskotts protokoll*, April 10, 1929, appendix 2).
31 Paulsson (1928), 11.
32 Paulsson (1928), 14; *Dagens Nyheter*, October 26, 1928; *Svenska Dagbladet*, October 26, 1928.
33 *Program och bestämmelser* (1928), 1.
34 Paulsson (1928), 11; (1930b), 35.
35 Paulsson (1930a), 23–4. Emphasis added.
36 Paulsson (1928), 12–13; (1937), 124–5.
37 Paulsson (1930b), 37–9.
38 Paulsson (1937), 225.
39 Paulsson (1937), 225.
40 Paulsson (1928). A more literal translation of this slogan would read: "The adapted to its purpose is the beautiful."
41 Cf. Ambjörnsson (1991b); Frykman (1981), 50–1; Hirdman (1990), 89–93; Rudberg (1981), 75. Some, but not all, Social Democratic ideologues regarded "radical architecture [as] an element in the current socialist struggle for reform" (Elin Wägner, author and sister-in-law of Gregor Paulsson, as quoted in Waldén [1980] 43). In l936, Per Albin Hansson, the Social Democratic prime minister, moved into a functionalist row-house.
42 Löfgren (1990b), 21.
43 *Svenska Dagbladet*, December 5, 1929.
44 *Säters Tidning*, May 16, 1930.
45 The here indicated bread and circus features are derived from hundreds of daily newspaper advertisements, exhibition advertising posters (SU-1930, box 237), and *Officiel huvudcatalog* (1930), 239–40, 243. Regarding spatially segregated amusement park zones at international expositions in general see Benedict (1983), 52–9.
46 The Stockholm Exhibition proved the most popular exposition held in Europe since 1925. After eleven weeks it had surpassed the total attendance of major exhibitions held in Barcelona and Antwerp (*Göteborgs Tidning*, August 15, 1930).
47 *Stockholms Tidningen*, May 15, 1930.
48 See Ståhle (1981, 145–6) regarding the probable origins of the *-is* endings so common in Stockholm slang usage.

49 "Eternit" was a composite of asbestos and cement.
50 See Pred (1991) for details.
51 Hughes (1991), 43, 42.
52 Cf. Rawson (1976), 253–4; Doorden (1988); Fuller (1988), 458; Etlin (1991). The "Rationalists sought a new and more rational synthesis between the nationalist values of Italian Classicism and ... the dynamic vocabulary of industrial form bequeathed to them by the Futurists." Before the movement was "undermined by the forces of cultural reaction" it could be claimed by a proponent that "Rationalist architecture was the only true expression of Fascist revolutionary principles" (Frampton, 1992, 203–4). Whatever the political influence and alignment of the Milanese Futurists, recent research strongly suggests they were not the main intellectual culprits in the rise of Fascism (Adamson, 1993).
53 Wrede (1980), 129. According to Wrede, Asplund went beyond the giants of the "Modern Movement" by placing his buildings in "an essentially traditional urban scheme" rather than dispersing them in a park-like setting.
54 Paulsson (1937), 33, 37. Since there was some division of labor into separate shifts, as many as 158 men were not on guard at the same time except when extra policemen were called on duty. Despite the millions attending the exhibition, only a very small number of legal transgressions occurred. As of July 26, only about thirty individuals had been arrested for drunkenness, shoplifting and other minor infractions (*Nya Dagligt Allehanda*, July 26, 1930).
55 On the territorialization of class and class-based experiences of Swedish urban spaces at this time see Frykman and Löfgren (1985) and Löfgren (1989c).
56 Näsström (1930b).
57 *Officiel huvudkatalog* (1930), 41.
58 *Arbetaren*, May 31, 1930. Because of the limited size of the buildings, the machinery shown was sometimes in model form rather than full scale.
59 The vendors – whose field of action was not restricted to The Corso – were so numerous as to prove pestering to many (*Svenska Dagbladet*, May 20, 1930; *Dagens Nyheter*, June 8, 1930).
60 Frykman (1981); Frykman and Löfgren (1985). The assertion that all social problems were rooted in poor hygiene and health, that germs were a threat to the nation as well as the family, had appeared since the 1890s as a part of reformist discourses elsewhere in Europe and North America (cf. Forty, 1986, 156–81).
61 *Verkställandeutskotts protokoll* (Minutes of the Executive Committee), April 5, 1930 (SU-1930 archive).
62 Diverse newspaper clippings, SU-1930 archive.
63 *Svenska Dagbladet*, May 27, 1930.
64 *Arbetet*, May 22, 1930; *Svenska Dagbladet*, June 6, 1930.
65 *Verkställandeutskotts protokoll*, May 23, 1930.
66 *Verkställandeutskotts protokoll*, June 19, 1930.
67 Physical attributes of the mast from Björklund (1967), vol. 2, 672; Karling

(1930), 40; Paulsson (1937), 91; *Dagens Nyheter*, October 20, 1930; *Stockholms Dagblad*, October 20, 1929; *Svenska Dagbladet*, October 18, 1929.
68   *Hallands Tidning*, May 16, 1930.
69   That the advertisers themselves quite likely had other designs, other intentions, is suggested by the following statement, made in Swedish Advertising (*Svensk Reklam*, October, 1930), shortly after the closing of the exhibition, when the first obvious signs of the depression were becoming apparent in Sweden. "It is not a matter for advertising to show the way to that disposal of citizen income which in the long run will prove most beneficial ethically and socially. Advertising can and should pull its share of the load, which it does by attempting to transform as much as possible of that income into actual purchasing power. Because purchasing power creates demand, which in turn leads to production, employment and job opportunities." (Quote from Björklund, (1967), vol. 1, 60.)
70   See Chapter 1, above.
71   Powell (1991), 6.
72   Edwards (1980), 32. For an extended treatment of the Russian constructivist school and its objectives see Loader (1983).
73   The slanting, rotating tower was to rise to a height of 1,300 feet. Hughes (1991, 92) has referred to it as "the most influential non-existent object of the twentieth century."
74   Ibid. Whether Asplund's mast was directly or indirectly inspired by constructivism is far from clear. In 1928 Asplund had been directly exposed to constructivist works at the Köln Presse Exhibition and at another exhibition in Brno. He also was acquainted with the Åbo (Turku) Exhibition of 1929, where the constructivist-influenced architecture of Aalto and Bryggman included several fifteen-meter advertising towers (*Svenska Dagbladet*, May 7, 1929; *Verkställandeutskotts protokoll*, August 23, 1929, appendix L). Another indirect source may have been the 1926 and 1927 works of Walter Gropius, Hannes Meyer and others in the Weimar Republic who had come under the sway of the Soviet avant-garde (Frampton, 1985, 35). Whatever the case, the following 1930 statement by a prominent constructivist (El Lissitzky) regarding the Vesnins' *Pravda* proposal, holds in every respect for Asplund's advertising mast: "All accessories which on a typical street are usually tacked onto a building – such as signs, advertising, clocks, loudspeakers and even the elevators inside – have been incorporated as integral elements of the design and combined into a unified whole. This is the aesthetic of constructivism." (*Russland – Architektur für eine Weltrevolution* as quoted in Frampton, 1985, 36.) Cf. Wrede (1980), 138.
75   Cf. Barthes (1979) and Bennett (1988, 97–8) on the Eiffel Tower, and previous chapter statements on the minarets of the Hall of Industry at the Stockholm Exhibition of 1897. The use of the mast/tower for publicity viewing purposes was made impossible by the three-person capacity of its elevator as well as the absence of an observation platform.

76 Cf. Walter Benjamin on urban illumination as summarized in Buck-Morss (1989, 308–9).
77 Karling (1930), 399.
78 Westerberg (1930), 340.
79 Paulsson himself claimed: "As an advertising form the exhibition is of import partly for the city and the nation, partly for the exhibitors" ("The Exhibition and Advertising," as reported in *Nya Dagligt Allehanda*, August 26, 1930).
80 Interview with Paulsson, *Nya Dagligt Allehanda*, December 10, 1928.
81 Interview with the exhibition's press chief, *Social-Demokraten*, September 14, 1928.
82 After a competition failed to yield any poster that met with the approval of Asplund and Paulsson, the task of design was assigned to Sigurd Lewerentz, a leading functionalist architect.
83 Special supplement on Sweden, *The Times*, of London, April 4, 1930.
84 Details regarding the advertising and publicity campaigns from: Björklund (1967), vol. 2, 669–71; *Dagens Nyheter*, July 24, 1930; *Nya Dagligt Allehanda*, July 25, 1930; Paulsson (1937), 49–60, 188, appendices 9 ("*Berättelse över pressarbetet* ... ") and 10 ("*Berättelse över reklamarbetet* ... "); *Preliminär plan för pressarbetet för konstindustriella utställningen i Stockholm år 1930* (SU-1930); *Verkställandeutskotts protokoll*, March 13, 1929, appendix 11; SU-1930 boxes 237–8 (*affischer*); *Svenska Dagbladet*, December 8, 1929.
85 *Stockholms Tidningen*, December 7, 1930.
86 SU-1930, clipping collection (a similar advertisement run on August 14–15 referred to 2,800,000 visitors). All preceding details from sources already indicated in note 84.
87 *Stockholms Dagblad*, July 3, 1930.
88 As part of the strategy of advertising the Swedish advertising industry, a special "Advertising Day" was organized. The event involved about 100 advertising executives from throughout Norden (*Dagens Nyheter*, August 26, 1930; *Nya Dagligt Allehanda*, August 26, 1930).
89 According to the contractual agreement between each exhibitor and the Executive Committee, advertising displays were "to be developed in consultation with the exhibition architect [Asplund] or his delegate," and any final proposal which was found "unsuitable" could be denied approval. (Contract copies appended to *Verkställandeutskotts protokoll*, November 12 and 29, 1929; January 31, April 25, May 2, 9, and 23, 1930.)
90 Asplund was strongly predisposed to approve such advertisements as he had come away from a 1928 tour of Bauhaus exhibitions and works convinced that "mobile features" were most likely to attract public attention (*Dagens Nyheter*, October 15, 1928). The poster-clad free bus which ran to the exhibition area from Stockholm's most fashionable department store (NK) was yet another form of advertisement in motion, as was the ill-fated mini-Zeppelin which flew the name of another department store.
91 Simple, spare typography had also "figured in de Stijl and Russian constructivist projects" (Wrede, 1980), 130.

92 *Officiel huvudkatalog* (1930); *Katalog över Bostadsavdelningen* (1930) and photographs from archival sources (ACM and SMA). General details in this paragraph are also from these sources as well as Björklund (1967), vol. 2, 673–5; Karling (1930); Råberg (1972), 340; and *Söndags-Nisse Strix*, July 2, 1930.
93 SU-1930, *Katalog och reklamtryck* (box 235).
94 *Stockholms Dagblad*, July 3, 1930.
95 *Verkställandeutskotts protokoll*, May 28, 1930.
96 *Stockholms Extrablad* (one-sheet daily newspaper distributed freely at the exhibition), May 29, 1930; *Svenska Dagbladet*, July 2, 1930; *Karlsborgs Tidning*, July 5, 1930.
97 Westerberg (1930), 340.
98 Anker Kirkeby, critic for the Danish newspaper *Politiken*, as translated in *Göteborgs Posten*, August 14, 1930.
99 Thorsten Laurin, outspoken opponent of functionalism, as quoted in *Svenska Dagbladet*, July 2, 1930.
100 *Stockholms Extrablad*, May 29, 1930.
101 *Svenska Dagbladet*, July 2, 1930; Paulsson (1937), 140.
102 *Social Demokraten*, August 4, 1930; *Nya Dagligt Allehanda*, August 4, 1930.
103 *Dagens Nyheter*, December 6, 1929.
104 *Stockholms Tidningen*, December 6, 1929.
105 Ibid.
106 Ibid; *Stockholms Dagblad*, June 26 and August 15, 1929; *Dagens Nyheter*, December 6, 1929. According to the secretary of the Association of National Advertisers such incidents were occurring daily.
107 *Stockholms Dagblad*, August 19, 1930.
108 *Stockholms Tidningen*, February 15, 1930.
109 *Stockholms Dagblad*, August 4, 1930.
110 *Nya Dagligt Allehanda*, July 16, 1929; *Stockholms Tidningen*, July 7, 1917.
111 *Social Demokraten*, August 15, 1929.
112 British architectural critic P. Morton Shand as quoted in Naylor (1990), 178.
113 Some contemporaries saw a resemblence to the *Haus Germania* in Berlin (*Nya Dagligt Allehanda*, June 9, 1930), while Frampton (1985, 36) has more recently made a case for constructivist inputs. Whatever the sources for Asplund's creation, his original scheme was somewhat less democratically intentioned, calling as it did for a little restaurant of high quality situated in the adjacent inlet on an Isle of Supreme Happiness, or Bliss (*Lycksalighetens ö*), that would be reached by "elegant water-taxis" (*Nya Dagligt Allehanda*, December 13, 1928). (The design actually employed for the exhibition as a whole was Asplund's third. Regarding the initial two proposals see Wrede, 1980, 127–9).
114 *Svenska Dagbladet*, April 3, 1930.
115 *Nya Dagligt Allehanda*, August 5, 1930.
116 ACM, miscellaneous holdings, box 44.
117 *Dagens Nyheter*, May 8, 1930. Both Paulsson and Asplund were themselves objectified, made subject to commodity fetishism, at *Gröna Udden*, an

exhibition café where one could drink a "Groggy Paul" or an "Asplundare" (*Norrlandsposten*, May 17, 1930).

118 For example, *Arbetet*, June 23, 1930; *Nerikes Tidningen*, June 6, 1930; and *Göteborgs Handelstidning*, May 20, 1930.
119 Reproduced in *Dagens Nyheter*, May 30, 1930 from *Söndags-Nissse Strix*.
120 *Stockholms Dagblad*, October 19, 1929.
121 Photographs in the holdings of ACM.
122 *Dagens Nyheter*, June 28, 1930.
123 *Svenska Dagbladet*, July 2, 1930; *Nya Dagligt Allehanda*, July 2, 1930.
124 *Folkets Dagblad Politiken*, August 15, 1930.
125 *Stockholms Dagblad*, August 22, 27, 1930.
126 *Verkställandeutskotts protokoll*, May 23, 1930; *Folkets Dagblad Politiken*, June 6, 1930.
127 *Dagens Nyheter*, September 28, 1930; *Social Demokraten*, September 28, 1930 (source of quote).
128 *Stockholms Dagblad*, September 26, 1930.
129 *Stockholms Dagblad*, October 17, 1930; *Nya Dagligt Allehanda*, November 12, 1930.
130 Herbert Grevenius in *Stockholms Dagblad*, June 4, 1930.
131 Nordström (1930a), 3. This same wording was employed by Bertil Nyström (1930) and Gregor Paulsson (1970, 127) who was also apparently involved in sketching the initial scheme for the contents of *Svea Rike* (Waldén, 1980), 51.
132 Nordström, as quoted in *Dagens Nyheter*, March 2, 1930.
133 Paulsson (1937), 168.
134 Nordström (1930b), 1. Emphasis added.
135 SU-1930, miscellaneous holdings, box 167.
136 Wahlund (1930), 177. The second motto loses much in translation, as it was presented with archaic spelling so as to emphasize its link with Swedish "tradition."
137 Löfgren (1991a), 101.
138 Ludvig Nordström as quoted in *Dagens Nyheter*, February 12, 1930.
139 *Lysekils Posten*, July 3, 1930.
140 Nordström (1930a), 43.
141 Neurath, as quoted in Galison (1990), 723.
142 Nordström (1930a), 3. Although Nordström is here referring to a book version of the *Svea Rike* exhibition, he and others employed the expression with respect to the exhibition itself. See Johansson (1930).
143 *Dagens Nyheter*, February 12, 1930.
144 Nordström (1930a), 7–8.
145 *Svenska Morgonbladet*, June 23, 1930. The article in question was subheaded "What Ludvig Nordström wants to teach the Swedes."
146 Nordström (1930a), 7.
147 Johansson (1930).
148 Nordström (1930a), 7.

149 Essence of a pictogram caption as quoted in *Dagens Nyheter*, March 3, 1930.
150 Nordström (1930a), 8–9.
151 *Verkställandeutskotts protokoll*, February 14, 1930. Paulsson and the other leading ideologues of functionalism regarded religious orthodoxy as characteristic of backward B-Europe (Asplund *et al.*, 1930, 17). In writings dating from after the exhibition, Lubbe Nordström "boiled over with wrath, loathing or contempt" for the unmodern attitudes of the State Church priesthood (Frykman, 1981, 42).
152 *Stockholms Tidningen*, June 12, 1930. State Church proponents were likely further miffed by the exhibition's inclusion of the evangelical movement among seven major turn-of-the-century grassroots movements depicted as "modern manifestations" of the age-old Swedish desire for "sunnier spiritual and material spaces" (Nordström, 1930a, 8).
153 Broberg and Tydén (1991), 52. A *norn* is one of the female Fates of Norse mythology.
154 Nordström (1930a), 21, 22, 24.
155 Nordström (1930a), 21.
156 Herman Lundborg, as interviewed in *Stockholms Tidningen*, March 5, 1930.
157 The development of Sweden's population politics and its accompanying discourses on "racial biology," "racial hygiene," eugenics and sterilization is treated at length in Broberg and Tydén (1991). Hirdman (1990, 92–158) situates "racial hygiene" within the modernization project of Alva and Gunnar Myrdal and other Social Democratic social engineers. Regarding these matters and the related politics of hygiene see Frykman (1981) and Frykman and Löfgren (1985). Also note Nilsson (1991).
158 Broberg and Tydén (1991), 37, 48–9. The Lundborg quote used by Broberg and Tydén is from his *Rasbiologi och rashygien* (1922). Lundborg claimed, among other things, that the "Jewish controlled press" was working against his institute.
159 The various Nazi parties in Sweden never received much over 50,000 votes during the 1930s and were therefore never represented in Parliament (Lööw, 1990, 264–8). However, elements of racial superiority and anti-semitism were present in the rhetoric of some parliamentary parties, especially that of the Farmers' Union, or *Bondeförbundet* (Broberg and Tydén, 1990, 54), which later in the decade formed a coalition government with the Social Democrats. Racism also appeared in a number of public forms, including a campaign carried out in the public schools in which milk drinking was represented by a "rational" white girl and coffee drinking by an "irrational" African boy (Frykman, 1981, 54–5).
160 *Stockholms Dagblad*, September 10, 1930.
161 *Dagens Nyheter*, July 7, 1930.
162 *Social Demokraten*, January 1, 1930.
163 *National Socialisten*, August 1, 1930; *Vidi*, July 3, 1930. *Vidi* was a magazine published by the Swedish Antisemitic Association (*Svenska Antisemitiska föreningen*).

164 *National Socialisten*, May 23, 1930.
165 *National Socialisten*, August 15, 1930.
166 *National Socialisten*, August 15 and June 6, 1930.
167 In popular usage, in the daily press and elsewhere outside of National Socialist circles, the wings-of-progress symbol was already referred to as "the razor," an instrument that was to clean things up and cut away all that was unnecessary (cf. Edwards, 1980, 20).
168 *Verställandeutskotts protokoll*, August 28, 1928 and May 26, 1930. Eliasson served as a convenient scapegoat, since the press had also more generally complained of scales, cash registers, typewriters and other foreign-made items being sold at exhibition-area shops. Paulsson countered that no such goods were to be found in the exhibition halls, that the shops had high rents to pay, and that shop rents were important to the exhibition's finances (*Svenska Dagbladet*, May 27, 1930).
169 Löfgren (1990a), 16. Cf. Anderson (1983) on nationalism and the construction of imagined communities.
170 That Swedes were not to carry their self-satisfaction too far was emphasized by a separate panel on which a caricature of Mr. Medelsvensson, or Mr. Average Swede, was offered as comic relief. The labels characterizing this chubby, well-feeling cartoon figure were at one and the same time charged with joking self-irony and instructive as to "new standards of morality" (Löfgren, 1990a, 16). Here was a figure whose idea of exercise was walking "the golden middle road;" a figure who regarded himself as a liberal, wanted others to think of him as a socialist, and was a royalist all the same; a figure who got along with everyone – especially before pub closing hours – but wished he was called something other than Medelsvensson. The meager, in-passing analyses of *Svea Rike* thus far generated by Swedish scholars have focused all too much on Medelsvensson and all too little on the remaining contents.
171 Nordström (1930a), 9.
172 Cf. Löfgren (1991a, 1991b) on the rhetoric of modernity in Sweden during the early 1930s. Also note Frykman (1981), 44.
173 *Höganäs Tidning*, July 26, 1930.
174 Samuelsson (1968), 232–3; Fürth (1979), 143, 189; Montgomery (1939), 240–1. Among those with bleak, rather than sunny, prospects were the watchmen, other unskilled exhibition employees and "extra" taxi drivers who had no idea where they would find work once the exhibition came to an end (*Nya Dagligt Allehanda*, September 10, 1930; *Dagens Nyheter*, September 14, 19, 1930). Those already unemployed in Stockholm were "in a desperate situation," experiencing a rapid decline in living standards despite some publicly provided unemployment relief (Fürth, 1979, 274). In retrospect the only accurately optimistic thing that could have been said about the future of the Swedish economy was that it would recover more rapidly from the "Great Depression" than the economies of most other countries, that better times would appear after the rock bottom winter of 1932–3.
175 *Folkets Dagblad Politiken*, January 29, 1930. The extensiveness of this campaign

is difficult to ascertain from this source, which wished to make political hay of the matter.
176 *Social Demokraten*, February 21, 1930.
177 *Dagens Nyheter*, September 6, 1929; *Social Demokraten*, September 6, 1929; *Nya Dagligt Allehanda*, September 16, 1929.
178 *Dagens Nyheter*, March 27, 1930; *Stockholms Dagblad*, March 29, 1930; *Stockholms Tidningen*, March 29, 1930; *Svenska Dagbladet*, May 2, 1930.
179 *Social Demokraten*, May 20, 1930; *Aftonbladet*, May 24, August 7, 1930; *Stockholms Dagblad*, May 26, 1930.
180 *Aftonbladet*, June 6, 1930.
181 *Social Demokraten*, June 4–6, 1930; *Stockholms Dagblad*, June 3, 1930; *Arbetaren*, June 4, 1930. The employer capitulated to all demands after more general action at the exhibition was threatened by the Hotel and Restaurant Workers' Union.
182 Paulsson (1937), 35–6; *Nya Dagligt Allehanda*, August 14, 1930; *Arbetaren*, August 14, 1930. Here too, after the Executive Committee interceded, the employer made concessions.
183 *Stockholms Dagblad*, July 2, 1930; *Folkets Dagblad Politiken*, July 3, 1930; *Nya Dagligt Allehanda*, October 10, 1930.
184 *Folkets Dagblad Politiken*, June 6, July 14, 1930. Vending within *Svea Rike* was permitted so long as it was conducted discreetly (SU-1930, miscellaneous holdings, box 167).
185 *Verkställandeutskotts protokoll*, August 21, November 14, 1930; *Nya Dagligt Allehanda*, August 29, 1930; *Stockholms Dagblad*, August 27, 1930; *Folkets Dagblad Politiken*, August 27, 1930.
186 *Nya Dagligt Allehanda*, July 7, 1930; *Social Demokraten*, July 8, 1930; *Stockholms Dagblad*, July 8, 1930.
187 *Aftonbladet*, July 15, 1930.
188 *Stockholms Dagblad*, July 20, 1930; *Arbetaren*, July 21, 1930; *Dagens Nyheter*, July 26, 1930; *Folkets Dagblad Politiken*, July 26, 1930; *Stockholms Tidningen*, July 27, 1930.
189 Reutersvärd (1988), 88. Unless where otherwise noted, this account is based on the writings of Oscar Reutersvärd (1980, 1988) and an interview (October 23, 1991) with that same distinguished artist and art historian. The exhibition in question also has been treated by Meregalli (1982, 141–68).
190 Brunius (1989), 183–4.
191 Carlsund as cited in Reutersvärd (1980, 57).
192 Carlsund (1930), 5.
193 Carlsund (1930), 5.
194 Carlsund as interviewed in *Stockholms Dagblad*, May 9, 1930.
195 Carlsund had previously sketched "*la rapidité cosmique*," an unsuccessful wall-painting proposal for the Einstein Observatory in Potsdam (Reutersvärd, 1988, 82).
196 As quoted in *Katrineholms Kuriren*, May 16, 1930.

197 Paulsson and others reponsible for the Stockholm Exhibition were from the outset uncertain, if not reluctant, about the inclusion of either an international or a solely Swedish art exhibit. Citing financial reasons, the Executive Committee had already rejected the possibility in 1928 (Kommittén, 1928a, 8; *Verkställandeutskotts protokoll*, June 20, July 3, 1928). Part of the reluctance, according to Reutersvärd (interview, October 23, 1991) stemmed from Paulsson's feeling that there was no Swede whose genius as a modern artist was comparable to that of Asplund's as an architect. Carlsund's proposal may have appeared especially attractive because it required no financial commitment on the part of the exhibition and yet allowed the event to promote the arts, just as most pre-World War II international expositions of any importance. Ley and Olds (1988, 199) note, for example, that the 1939 New York World's Fair included avant-garde art even though it was "described as the consumer's fair, and promoted widespread access to consumer goods and services."
198 Reutersvärd (1988), 89.
199 Reutersvärd (1990).
200 Näsström (1930a).
201 *Svenska Morgonbladet*, August 19, 1930.
202 *Stockholms Tidningen*, August 25, 1930.
203 *Dagens Nyheter*, August 20, 1930.
204 *Stockholms Tidningen*, August 19 (?), 1930, as quoted in Reutersvärd (1988, 92).
205 *Svenska Dagbladet*, August 22, 1930.
206 Johnson (1961), 45.
207 Carlsund as quoted by Reutersvärd (interview, October 23, 1991).
208 Reutersvärd (1988), 95, 93.
209 All quotes from Reutersvärd (1988), 98.
210 Reutersvärd (interview, October 23, 1991).
211 Näsström (1930a).
212 Waldén (1980), 38.
213 Ekbom (1991) reviewing Segal (1987).
214 Löfgren (1991b), 3.
215 Cf. Löfgren (1990b), 30.
216 *Verkställandeutskotts protokoll*, August 8, 1930.
217 *Göteborgs Posten*, August 14, 1930.
218 Cartoon from *Stockholms Tidningen*, May 11, 1930. Cf. comments by Löfgren (1990b, 21 and 1990c, 190).
219 *Svensk Skrädderitidning*, no. 14, 1930. The cartoon – which was captioned "The undisguised joy of the old furniture at the arrival of the newcomer" – actually hung for a while at the Humorists' Parlor.
220 *Stockholms Dagblad*, January 8, 1931.
221 *Stockholms Dagblad*, June 19, 1930.
222 Letter to the editor, *Svenska Dagbladet*, July 5, 1930.
223 Paulsson (1937), 65–6. Even if one allows for a very large number of people making multiple visits, these figures must be regarded as astonishingly low.

224 During the first three days of the exhibition alone there were forty-two accidents, most commonly involving glass-wall collisions (*Nya Dagligt Allehanda*, May 20, 1930).
225 Gaunt (1983) as cited in Löfgren (1990b), 21.
226 Cf. Löfgren (1990b), 21 and (1990c), 191.
227 Tarschys (1991).
228 H. L. (1930); Svedberg (1980), 62. Cf. Hirdman (1990), 164. Such revolutionary ideas were not without precedent. For example, as early as 1916, Alice Constance Austin, a self-taught architect, had proposed kitchenless houses and meal deliveries from a central kitchen as part of her proposed feminist socialist city for 10,000 residents in California. For details on this and other radical housing schemes developed by U.S. feminists see Hayden (1981).
229 Cf. Råberg (1972); Waldén (1980).
230 Frykman and Löfgren (1987), 149; Hirdman (1990), 104.
231 Lind (1991). The conversion of modern apartment living rooms into traditional working-class "fine rooms," into spaces to be used only on Christmas Eve and other highly special occasions, remained a common practice well into the 1940s (cf. Unge, 1992, 19).
232 Westerberg (1930), 343.
233 *Göteborgs Posten*, August 14, 1930.
234 *Dagens Nyheter*, July 16, September 22, 1930; *Social Demokraten*, July 18, 1930; *Stockholms Dagblad*, July 17, August 5, September 1, 1930; *Stockholms Tidningen*, July 18, September 25, 1930; *Svenska Dagbladet*, July 17, September 23, 1930.
235 *Aftonbladet*, September 3, 1930; *Stockholms Dagblad*, September 10, 1930; *Svenska Dagbladet*, September 3, 1930.
236 *Stockholms Tidningen*, May 15, 1930. The early appearance of these usages most likely stems from the fact that tens of thousands of Stockholmers had taken Sunday strolls through the exhibition area while its buildings were still under various stages of construction.
237 *Social Demokraten*, May 20, 1930; *Östergötlands Dagblad*, May 22, 1930.
238 *Kaspar*, July, 1930. Cf. note 114 above.
239 *Allt för Alla*, May 18, 1930.
240 Cf. Willis (1990) on the symbolic creativity of common culture, the symbolic creativity associated with seemingly mundane activities.
241 Cf. pp. 139–40, above.
242 Slang terms from Björklund (1967), vol. 2, 672.
243 *Stockholms Dagblad*, December 12–16, 1928; *Järnhandlaren*, April 1, 1930.
244 *Dagens Nyheter* (September 30, 1930) and other newspapers ran stories about the plundering "orgy" of 1897, raising the specter of a repeat occurrence. Additional guards were stationed at *Svea Rike* and other attractions to prevent "an all too lively souvenir collecting" (*Stockholms Tidningen*, August 23, 1930) and the police assigned an extra contingent consisting of "every available man" (*Dagens Nyheter*, September 30, 1930). Despite the presence of almost 80,000 people on the final evening, there were no more than a few minor incidents: a

gang of youths stole some signs; a young woman made off with some flowers; two men snuck away with a stage prop from one of the amusement-park theaters; and three well dressed people removed a scale from The Paradise, only to return it the next morning (*Göteborgs Tidningen*, October 1, 1930; *Stockholms Dagblad*, October 1, 1930; *Nya Dagligt Allehanda*, October 1–2, 1930).

245 Waldén (1980), 46.
246 Frykman (1981), 37.
247 Cf. Frykman (1981), 59–60.
248 *Folkets Dagblad Politiken*, August 5, 1930.
249 Huyssen (1986), 11. Cf. Asplund *et al.* (1931).
250 This watercolor and india ink work was a "reworked version of the artist's admired 1924 attempt to realize Léger's central idea of harmonizing the opposition between the technological and human form-worlds" (Reutersvärd, 1988, 108). It was part of his "*adieu* to all the ideals of *Art Concret*" and a shift toward the "irrational, emotionally laden" surrealism he once publicly abhorred (Reutersvärd, 1980, 57).
251 Cf. Olsson (1990, 1991a, 1991b) regarding "invisible maps" in general, and their relation to the power exercised by the bureaucracy of the modern Swedish State in particular.
252 Cf. the arguments by Fürth (1979) regarding the changes "in mental outlook," the breach with past Swedish developments, that came in the wake of the depression experience.
253 Löfgren (1991b), 11; Unge (1992), 28.

CHAPTER THREE

WHERE IN THE WORLD AM I, ARE WE? WHO IN THE WORLD AM I, ARE WE?: THE GLOB(E)ALIZATION OF STOCKHOLM, SWEDEN

•

THE GLOBE ...
A FEELING OF COMMUNITY THAT SPANS ALL BOUNDARIES

Globalization does not signal the erasure of local difference, but in a strange way its converse; it revalidates and reconstitutes place, locality and difference.

MICHAEL J. WATTS
(1991), 10

Since the end of the seventies a new artistic and intellectual culture with official support and with public success has emerged which I call citadel culture. This culture has freed itself from any direct political reference to the present times with all the more self-assurance the more threatening contemporary history appears.... Citadel culture is deployed without limitation, expresses itself without inhibitions, is capriciously designed and grandiosely staged.

O.K. WERCKMEISTER
(1991), 12–13

In the space of just a few short years, Stockholm has acquired a new landmark and Sweden a new national arena – The Stockholm Globe Arena. An arena that has very quickly become the obvious goal for the world's leading athletes, entertainers and cultural personalities. Only the Globe can provide the backdrop necessary for those truly giant events. And only in the Globe can 16,000 people meet and enjoy a feeling of community which spans all boundaries.

The Globe Arena is Sweden's first world-class arena. It is modelled on international standards and is equipped with private boxes, restaurants and an internal TV network. Thanks to its tremendous resources and unique shape, the Globe has awakened the world's interest in Sweden and Stockholm as the meeting place for world-class events.

AB Stockholm Globe Arena
(1989), 3

The Globe is one of the largest, most interesting and most advanced building projects in Sweden of the twentieth century. The Globe is the world's biggest spherical building... [N]othing like the Globe has ever been built anywhere in the world.

AB Stockholm Globe Arena (1989), 1

The Globe is a typical product of the information- and consumption-society.... With the help of television the image of "Stockholm's Eiffel Tower" will be trumpeted about. The image of an international, dynamic power-center. "World-class" is one of the key words surrounding all projects sponsored by Stockholm authorities, including the Globe. That events in the Globe will be entirely oriented toward the spectacular is given.... The Globe is a symbol of our times.

TORBJÖRN ANDERSSON et al. (1991), 7

Power has always built monuments. In order to mark. To impress. To frighten and intimidate.

In times past pyramids served the purpose. Now Power has taken a new form and become more global.

With the Globe Sweden reassumes a place in the proud tradition of monument builders. We haven't built anything so grandly scaled and pompous since the long gone Great Power Era.[1]

The Globe is a confirmation of the here and now: of Sweden in the economically overheated, quick-deal, quick-profit '80s. The epoch of financial whizkids. Stockholm in the age of one stock-market increase heaped upon another. A globe built in the radiance of an [extended] buying spree.

A party cathedral for the spectacle of the times. Built for the gladiator games of the twenty-first century. Built for the Pope, Tre Kronor [the national hockey team], and the Boss.... [B]uilt in order to radiate success through an unholy alliance of business and construction firms, banking capital, high technology and political power holders.

EVA GÖRANSSON (1989)

[M]ore than ever, the urban landscape relies on image consumption.

[Urban landscapes] always show asymmetries of power.

SHARON ZUKIN (1991), 38, 274

It was a spectacle of citadel culture. This is the metaphor I use here to characterize the dominant artistic and intellectual culture of the democratic industrial societies during the years 1980–87, the time of their greatest economic success, a culture contrived to exhibit the conflicts of those societies in a form that keeps any judgment in abeyance.

O.K. WERCKMEISTER (1991), 3–4

GREIL MARCUS
(1989), 99²

"The spectacle," Debord said, was "capital accumulated until it becomes an image." A never-ending accumulation of spectacles – advertisements, entertainments, traffic, skyscrapers, political campaigns, department stores, sports events, newscasts, art tours, foreign wars, space launchings – made a modern world [globe], a world in which all communication flowed in one direction, from the powerful to the powerless. One could not respond, or talk back, or intervene, but one did not want to. In the spectacle, passivity was simultaneously the means and the end of a great hidden project, a project of social control. On the terms of its particular form of hegemony the spectacle naturally produced not actors but spectators: modern men and women, the citizens of the most advanced societies on earth, who were thrilled to watch whatever it was they were given to watch.

MICHEL DE CERTEAU
(1984), xiii, xvii

The presence and circulation of a representation ... tell us nothing about what it is for its users. We must first analyze its manipulation by users who are not its makers. Only then can we gauge the difference or similarity between the production of the image and the secondary production hidden in the process of its utilization. ... The tactics of consumption, the ingenious ways in which the weak make use of the strong, ... lend a political dimension to everyday practices.

BO LUNDGREN
[member of the Moderate, or Conservative, Party and Minister of Taxation], (1992)

The aim of [our] tax policy is, and remains to be, a successive reduction of tax levels in order to fulfill the overlapping objectives of greater individual freedom [to consume] and a higher rate of growth.

GÖRAN TUNHAMMAR
[President of SAF, or the Swedish Employers' Confederation, Sweden's most powerful business organization], (1992)

The economic significance of European Community membership is not easily made clear. ... [For] it is a group completely absent from the Swedish debate, namely the consumers, who are the big material winners from integration into the European Community. The thought behind integration is to make it better for consumers, not producers. Well-functioning competition within a very large, free domestic market is the method.

KRZYSZTOF POMIAN
as interviewed in Otto Mannheimer (1992)

Maastricht is not merely a technical problem, but also a deep-going political one. The agreement appears to have been made rather hastily. ... [T]he threat to personal identity felt by many Europeans today was [not] taken into consideration.

> [Cultural identity] is not something which already exists, transcending place, time, history and culture.... Far from being eternally fixed in some essentialized past, [cultural identities] are subject to the continual play of history, culture and power.

STUART HALL (1989), 70 and (1991), 25

> The erosion of the nation-state, national economies and national cultural identities is a very complex and dangerous moment.
>
> Now more than ever, and in the future even more than now, everybody's personal identity, and every collective hope, depend on complex mediation between the local and the global.

MARSHALL BERMAN, on jacket to Zukin (1991)

> *I have lost my identity.*

Construction worker from near the town of Kramfors, as interviewed on the Swedish national television news program *Aktuellt*, August 16, 1991, when he had been unemployed for eight months

---

> *The [Globe] arena will become a house of possibilities, a house for everybody; yes, a real all-activity house.*

STEFAN HOLMGREN [President of AB Stockholm Globe Arena], (1987 statements as quoted in Torbjörn Andersson *et al.*, 1991, 4)

> *Our little city continues to struggle and pant in order to get its pulse to beat in time with the world metropolises. And the more hysterically it expands and attempts to harmonize the more Stockholm appears like a country cousin.*

TOMAS ANDERSSON (1992)

The Globe, like the political and economic world at large, like global forms of capital, repeatedly appears out of nowhere in Stockholm. Just when it is completely removed from thought, it springs into sight, familiar from last time, but somehow different – now seen from a different angle, in a new light. A visual jolt. A shock of (hyper)modernity repeatedly encountered when navigating the city's streets, in the midst of everyday life, in the course of conducting ordinary and extraordinary tasks, in the flow of daily experience; straight ahead in moments of noise-drenched public activity; over the shoulders of another, through an apartment window, in a moment of whispering intimacy.

The Globe, "rising like a ghostly presence from the freeway net of the city,"³ like the political and economic world at large, like global forms of capital, suddenly looms unanticipated on the Stockholm horizon. Arresting attention. Stunning. Dazing. Disturbing. Interrupting. Interfering. Unexpectedly announcing its domination. Intruding into the field of awareness yet eluding anything more than partial comprehension. Dwarfing, yet distant. Reminding of a presence that is always there, inescapable, if not always perceived. A structure of ubiquitous reach. Often appealing, overwhelmingly graceful on first glance. Yet, often threatening to bowl over, to knock down, to unmoor, to disorient, to dislocate, to unposition with its heavy immensity.

The Globe, like the political and economic world at large, like global forms of capital, capable of evoking a sense of disjuncture, of discontinuity, of incongruity, of rupture; capable, by way of its repeated unannounced appearances, of leaving woman, man and collectivity in wondering puzzlement.

What in the world is going on (t)here?
Where in the world am I, are we?
Who in the world am I, are we?

The Globe, a spectacular space for commodified entertainment spectacles, for the consumption of commodified bodies, for the display of commodities, for megaceremonies of pomp and circumstance – such as the 1991 Nobel Prize festivities – designed for national and global consumption, for large-scale meetings where stockholders are given accounts of rising or falling revenues, of rising or falling levels of consumption. A spectacular space culminating a century of modern consumption, of consumption under conditions of industrial modernity, high modernity and hypermodernity. The Globe, a spectacular space of consumption in a place and era where the consumption of spectacles had been reduced to an everyday matter.

The Globe, a spectacular polysemous structure. A spectacular monument of multiple meanings, of meanings projected and rejected, of meanings from above meant to work idea-logically, of meanings frequently reworked from below. A spectacular concretization of (would-be) hegemonic discourses, of hegemonic discourses that are not always consumed without symbolic contestation. A spectacular house of play, a "house for everybody," a house where "a feeling of community" is to "span all boundaries," a house to promote collective content. And yet, a house that engenders symbolic discontent, that precipitates the cultural play of difference, that gives rise to cultural politics which demote, belittle, scale down to size. A spectacular, all-consuming material structure through and in which metaphors and metonyms of current centrality are materialized. Metaphors and metonyms pertaining to, among other things, the global power of capital, the increasingly free reign of market forces in Sweden, and the concomitant crisis of the welfare state. Pertaining to the articulation of the local and the global, to the position of Stockholm (and Sweden as a whole) in the network of globally circulating capital. Pertaining to the various positions taken by Stockholmers (and Swedes in general) *vis-à-vis* the economic and political world, the possibility of European Community membership and the global situation. Pertaining directly or indirectly, in most instances, to the profound individual and collective identity "crises" abounding in Sweden.

GLOB(E)AL CONTEXTS

The Globe did not arise out of a vacuum. Its construction fits within a set of geographically specific architectural (hi)stories, including Gunnar Asplund's planetarium at the Stockholm Exhibition of 1930. It had a not always flattering array of antecedents, a number of predecessors of sometimes disturbing design, an assortment of formal precursors and contemporaries with which it was (and is) frequently in eerie resonance.

The first proposals for globe-shaped buildings arose in the era of the

French Revolution.[4] Architects committed to the causes and discourses of the revolution saw their task as that of "devising radically new types of public buildings for a radically reshaped society with radically new collective functions." Their "utopian dream images" ranged from floorless ball-shaped interior rooms, to spheres of "colossal format."[5] Ledoux not only offered a globular multi-storied residence for the employees of an estate, but a fully spherical columbarium, a sepulcher-temple for the cremated. Lequeux drew up a spherical temple for worship of the planet earth and "the holy equality." The most grandly scaled and dramatic of these globe-shaped proposals were Etienne-Louis Boulée's three plans for a temple to Isaac Newton.[6] ... Time and time again the globe as temple. The Globe like its revolutionary forebears as pure-of-line structure poised to "lift from the ground," as simple-lined structure "drawn toward the heavens and the cosmos,"[7] as temple? The Globe as temple to spectacular consumption?

During the first half of the nineteenth century at least three terrestrial globes were constructed for educational ends. The *Géoramas* of Delanylard and Charles Guérin (Paris, 1823 and 1844 respectively) were apparently designed to yield profits as well as enlightenment regarding the earth's physical features; while the globe at Philippe Vandermaelen's *Éstablissement Géographique* (Brussels, 1830) was more strictly instructional in purpose.[8] ... Here and there the globe as teaching device. The Globe as edifying edifice? The Globe as edifice for teaching the lessons of spectacular (hyper)modern consumption?

Commencing with the first world's fair, London's Great Exhibition of 1851, globular or globe-like constructions periodically have appeared at international expositions. Wyld's Great Globe was a brick rotunda within which vistors "mounted a series of staircases to view plaster casts of the world's continents and oceans."[9] Its successors have included the Villard-Cotard globe displayed at the 1889 Paris *Exposition Universelle*;[10] the cannon-ball shaped pavilion of a French munitions foundry at the 1900 Paris *Exposition Universelle*; the 60-meter-high Perisphere which (together with an adjacent triangular obelisk, the Trylon) dramatically symbolized the 1939 New York World's Fair; the Unisphere of the 1964 New York World's Fair, a huge, if not somewhat "tacky,"[11] globe properly tilted at 23.5 degrees; and Buckminster Fuller's geodesic dome which served as the U.S. pavilion at the 1967 Montreal World's Fair. In most of these instances the structure was a synecdochical statement for the exposition's spectacular space as a whole. A synecdochical statement rendering – along with the representation of foreign goods and the recreation of "exotic" or colonial street and village scenes – "the whole world metonymically present." A synecdochical statement which attempted to help order the consumer objects and peoples of the world into a single knowable totality; which attempted to regulate the now and future vision of the spectator by rendering "the whole world ... subordinate to the controlling vision of the spectator," "to the dominating gaze of the white, bourgeois, ... and male eye of the metropolitan powers."[12] A synecdochical statement which projected a view of the world in keeping with hegemonic interests, in keeping with would-be hegemonic messages regarding progress, national pride and racial superiority, and with would-be hegemonic messages that legitimated the world as it was.... Over and over again, the globe as illusory device helping to show the public at large that a seemingly chaotic, incomprehensible world is not so, that the world of the host nation is a good place getting better through the world of consumption. The Globe as such an all-the-world-in-one-place illusory device? The Globe as would-be hegemonic device where domination and legitimation are marketed through placing the world of (entertainment) consumption at the feet of the spectator?

Especially during the 1930s, material representations of the globe were deployed as a promotional image, as a means of underlining the scope of corporate activities, as a symbolization of the reach and significance of goods and services offered, as a statement of dependability and success, of having conquered the world. In Prague a utilities company built a large spherical gas-holder, in New York the *Daily News* set a 12-foot globe outside its offices, while in Miami and Detroit somewhat bigger globes were created for Pan American Airways and the Ford Motor Company.[13] (More recently in Atlanta, a 50-foot globe emblazoned with the Coca-Cola trademark has been suspended from a lattice-work overhang in front of "The World of Coca Cola" a museum of "Coca Cola history and memorabilia.") In city after city, the globe as consumer-attracting sign of achievement, as show(-off)piece of capital. The Globe as spectacular success signifier, as attention-grabbing commercial image, as show(-off)place of global entertainment capital?

As the centerpiece of his design for the massive renovation of Third-Reich Berlin, Albert Speer planned a "People's Hall," nothing less than the world's largest building, a globe-like Pantheon for the glorification of Hitler and the state, to be situated at the end of a monumental avenue. This structure, like other neoclassical Nazi state buildings, was intended to "reawaken" the soul of the German people.[14] ... In a fearsomely dark then and there, the globe as state glorifier, as ideological force. In a not exactly bright here and now, the Globe as monument to the local Stockholm state? The Globe as consumption dreamworld provider, and therefore as lulling, rather than awakening, ideological force?

At the Disney Epcot Center near Orlando, Florida, a 60-meter high geodesic sphere, "Spaceship Earth," a "totem of universality," serves as the symbol for a theme park whose "main street is flanked by the pavilions of

major U.S. corporations, each housing some version of a 'ride' through a halcyon future."[15] (In commercials for the theme park, Mickey Mouse waves from atop the sphere.) The *La Vilette* district of Paris, a park including science and industry exhibits, houses *La Géode*, "a glistening 36-meter-diameter steel-covered sphere ... in which the largest film screen in the world displays travel in the universe, dives into the ocean and flights between feather-light clouds."[16] Elsewhere in France, at a similar park in Poitiers, a shiny "Futurescope" globe rests upon a prism-shaped building suggesting a sunrise, the birth of "a new world in which education, leisure, production, film and science" fruitfully interact. Here "visitors take a telescopic elevator" to the center of the sphere and then "travel further between atoms and galaxies at the speed of light."[17] In Toronto, a similarly dimensioned globular building functions as a theater at the Ontario Place entertainment center. At the Shonandai Cultural and Social Center in Fujisawa, Japan, there are two aluminum-covered spherical buildings; one containing an auditorium for various types of performance, and the other – covered with a map of the world – enveloping a planetarium.... Around the entire industrialized world, a set of globes providing an escape from the troubles and tensions of everyday life, a set of globes of escape and sedation where future wonders are presented in an effort to erase present woes, or where planetary or theatrical other-worlds are offered as a distraction from the pressures and frictions of the lived world. The Globe as commodified escape, as therapeutic release from present-day Stockholm? The Globe as tranquillizing escape from the cares and conflicts of here-and-now Stockholm?

The Globe did not arise out of a vacuum. It was conceived, politically maneuvered, financed and constructed in a highly volatile world of coexisting and globally integrated capitalisms, in a world of speeded up instability and incessant economic redifferentiation, in a world of difference, connectedness and structure repeatedly transformed.[18] A world in which representatives of Stockholm-based Swedish capital and the local Stockholm state were attempting to (re)position themselves, to (re)assert themselves globally, to project a new image.

The Globe came into being in, and was a contributing response to, a world of unrelenting and accelerated flux. A world where the "inconstant geography of capitalism"[19] was in constant evidence. A world in which new multinational patterns of production were being made and unmade within sector after sector, not infrequently with consequences for Stockholm-based corporations and financial institutions. A world in which manufacturing production systems were not only becoming more internationalized, but relentlessly reconstituted and deconstituted, either through episodes of investment and disinvestment, the widespread introduction of more "flexible" and decentralized forms of interplant and interfirm organization, or through the alteration of input-purchasing arrangements. A world in which industrial production systems were relentlessly reconstituted and deconstituted in ways that on occasion after occasion directly affected operations at units located within metropolitan Stockholm itself or its extensive domestic hinterland.

The Globe came into being in a world in which the constant churning of the international division of labor, the tireless abandonment of capital for new manufacturing facilities in low-labor-cost Third-World countries, and the ceaseless movement of new-plant capital from one market-rich OECD country to another, did not always fail to touch Stockholm either directly or indirectly, via other places in Sweden. A world in which neither industry-particular crises of overproduction, overcapacity and overaccumulation, nor the unremitting shift of resources from activities with falling rates of return to activities which seemed to promise higher return rates, were without occasional repercussions for Stockholm employers and employees. A world in which neither the consequent geographically specific instances of creative destruction and environmental damage, nor the accompanying deindustrialization and economic restructuring of one region after another, were without their impacts on Stockholm. A world in which the indefatigable expansion of capitalist relations of production into new Asian, African and Latin American locations did not always leave Stockholm corporations on the outside looking in. A world in which neither the passing of once viable production technologies into obsolescence, nor the easing of corporate control at a distance through the further improvement of communications and transportation technologies, nor the generation of enormous migration streams by economic and political turmoil, could always leave Stockholm unmarked by direct involvement or second-order ramifications. A world in which the ownership patterns of multinational corporations – including those with headquarters or otherwise operating in Stockholm – were subject to constant reshuffling.

The Globe came into being in a world in which money capital had become hypermobile, in which national financial capital markets were becoming increasingly intertangled with one another through the massive electronic transfer of funds and other means, in which "'a new international financial system' devoted to the creation of credit and debt . . . outside national control" was rapidly emerging,[20] thereby forcing repeated alterations and realterations both in the behavior of Stockholm's major banks and in the role of the city as a secondary international financial center. A world in which the multiplication of new forms of international subcontracting[21] and franchising was coming to mean far more to Stockholm than the McDonaldization of its urban landscape. A world in which the standard-of-living gap between countries of the "North" and "South" was being yet further magnified,[22] in which the everyday life of most OECD residents was becoming ever more commodified, ever more awash in new and highly differentiated consumer goods and services. A world in which the intrusion of foreign goods into local markets and lives was becoming intensified throughout North America and Western Europe, in which the middle- and upper-class inhabitants of those parts of the world – including most Stockholmers – found themselves all the more "able to draw from a . . . [wide] repertoire of instantly accessible symbolic goods and styles from the 'global showcase.'"[23] A world in which satellite television was further facilitating the global linkage of mass-culture and personal consumption markets, in which entertainment, advertising and "news" images now "much more rapidly and more easily" crossed and recrossed "linguistic frontiers,"[24] thereby further contributing to a perpetual reshaping of the ways in which different national capitalisms interpenetrated with one another in any given location.

The Globe came into being in a world in which all of these phenomena and relations were in metamorphosizing turmoil at once. These phenomena and relations were, furthermore, synonymous with the emergence of new major metropolitan centers of accumulation within the global capitalist economy; with an up- and down-surge in the relative position of centers of longer-standing primary and secondary importance; with an instability in the relative and absolute significance of metropolitan nodes within global networks of finance, production, exchange and corporate administration;[25] with new metropolitan challengers to the traditional dominance of Paris and New York within the "fashion, culture and entertainment industries, [as well as] television, publishing and music."[26] This being the case, these phenomena and relations were also synonymous with with new forms of economic competition among the large metropolitan complexes in question. The economic and political leadership of those centers were now striving for, among other things: new ways of marketing themselves through the mobilization of culture; new forms of international visibility; new forms of positioning investment; new concrete outlets through which overaccumulated capital could be deployed to create symbolic capital, to project a new global image, a new social representation of place – one capable of alluring economic activity, of generating more local jobs, higher property values and greater retail sales.[27]

The Globe came into being in a Stockholm that was part of a global system of capitalist metropolises continually subject to transformative upheaval; in a Stockholm, moreover, whose future investment climate was becoming clouded by uncertainties regarding Sweden's status *vis-à-vis* a European Community that was moving toward greater integration – by mounting fears of the consequences of being an indirectly involved bystander, rather than an active participant. In a period of momentous global and continental transformations, was Stockholm to maintain its position as a metropolis of major secondary importance?

Was it to enhance its role in a world of (con)fusing capitalisms?

Or was it to be peripheralized into insignificance?

GLOB(E)AL POLITICS

The steady growth of citadel culture in scope and funding seems to have kept pace with the rebound of the capitalist economy after the crisis of 1980–81, in correspondence with the general shift of priorities from [social-welfare oriented] political programs ... to market-oriented investments. While state expenditures for educational programs are shrinking, while budgets and payrolls for adult education and libraries for research institutions and universities are being cut, a

O.K. WERCKMEISTER (1991), 15

spectacular culture is being funded generously by variable combinations of state subsidies, corporate investments, advertising accounts, and returns from large audiences.

DAVID HARVEY
(1989b), 271, 303

The free flow of capital across the surface of the globe ... places strong emphasis upon the particular qualities of the spaces to which the capital might be attracted. The shrinkage of space that brings diverse communities across the globe into competition with each other implies localized competitive strategies and a heightened sense of awareness of what makes a place special and gives it a competitive advantage.

Capitalist hegemony over space puts the aesthetics of place very much back on the agenda.

STIG ANESÄTER
(1989)

Uncontrolled market forces have been given a freedom of movement which isn't confined to their old playground, ... [having] also infiltrated the public sector, not entirely without support from a smiling business community.... The Globe is [one] example of the laxity resulting when the market economy is allowed to assert itself at the taxpayers' expense.

THOMAS HALL
(1989)

[Given the background of its anticipated long-term influence on the region], one might expect the [Globe] project was preceded by careful planning, where consequences and alternatives were deeply explored and made public. Such is not the case. The Globe is not the result of any public planning worthy of the name, but of negotiations between a handful of municipal politicians and those firms responsible for its construction and financing.

Liberal Party member of the Stockholm Municipal Council, as quoted in Sahlin-Andersson (1989), 200

The matter was forced through in an unparalleled manner. Hardly a moment after statistics were presented, one was expected to take a position.

The Tale of the Little Ice Stadium, a satirical appendix to TORBJÖRN ANDERSSON et al. (1991), 17

And think what luck one can have. One of those [Municipal Council members] who was involved in deciding happened in fact to be the head of a large construction firm. His name was Hans. And in order that Hans' firm would be able to build the satellite project a competition was arranged which Hans' firm was allowed to win. And that even though Hans, in fact, left the room just when they were going to vote.

Part of the area where the Globe now stands had been a parking lot, while the remainder had for decades functioned as a play-space. The latter contained a small wooded area with two "lizard ponds" where, in the absence of supervision, children imaginatively created their own exciting worlds, swam, caught salamanders, fished and, according to season, wildly rode either their bicycles or their sleds and skates. It also encompassed a set of municipally owned soccer fields primarily serving men and boys associated either with the nearby slaughterhouses and stockyards, or with the working-class residential area within Stockholm's *Södermalm* district just to the north.[28] Adjacent to the site was the *Johanneshov* arena, a building which had gradually been transformed from a simple skating rink into Stockholm's premier facility for ice-hockey matches. During the 1970s, however, it became all the more evident that *Johanneshov* was too small and age-scarred either to accommodate a world-championship hockey tournament, or to attract the variety of sporting and entertainment events then making their international-circuit stops at the *Scandinavium*, a spacious modern structure situated in – horror of all horrors – Göteborg, Sweden's second city.

In this vacuum a number of new-arena proposals were put forward by major construction companies and other interested parties in conjunction with plans either to develop entirely new suburban areas, or to renovate certain parts of the central city. All such schemes ran out in the sand until the early 1980s when members of the municipal council, apparently picking up on conversations among powerful city administrators, became actively engaged in the matter, soon focusing on the possibilities of an arena on land beside the now discredited *Johanneshov*. Here, as on other occasions involving investment and development politics in Stockholm and the nation as a whole, an unholy alliance was formed between Social Democrats and members of the right-wing Moderate Party; more precisely, especially between prime mover Ingemar Josefsson, Social Democratic head of the council's Leisure and Cultural Activity Commission, and Hans Liljeroth, a Moderate councilman, who simultaneously was the top Stockholm executive for SIAB, one of Sweden's largest construction firms.[29]

It was clear from the outset that the costs of a new arena were far beyond the capacity of the municipal budget. This was especially so given the competing call for municipal funds to alleviate a growing housing shortage, precipitated by an expanding population and the real-estate pressures created by the city's burgeoning financial and service sectors.[30] If, as authorities desired, a sports and cultural-events complex "unique for Norden"[31] actually was to be constructed, then it would require some concession to "market forces," some accommodation of pent-up demand, some kind of co-operation with private capital, some kind of collaboration with those almost frantically speculating in office-space and real-estate development. In short, the project would require some kind of arrangement similar to that recently worked out for the simultaneous renovation of the city's Central Station and the construction of, among other things, a new bus terminal and office space in a self-consciously postmodern building above the station's previously exposed railroad tracks. With the knowledge that swift action was necessary if Stockholm was to host a world hockey championship before the end of the decade, in July 1984 some thirty large firms were approached to determine whether they were interested in a project competition not only for a new arena, but the entire *Johanneshov* area – a competition which, because of the ambitious scale involved, would require the formation of a joint-venture concern. After some subsequent council debate, a design- and financing-proposal competition for the arena plus 75–90,000 square meters of commercial space (for offices, a hotel and shops) was announced on March 1, 1985, and it was eventually agreed that the municipality would budget no more than 26 million crowns for the scheme.[32]

Slightly more than a year later, following extended off-the-record financial and planning negotiations, the project was awarded to the Court (*Hovet*) Consortium, a group of eight companies, including Hans Liljeroth's SIAB, whose architectural proposal was highlighted by a globular arena rather than the circular building with a cupola roof suggested in the competition program. (During the initial phases of negotiation Liljeroth had managed to act as *Hovet's* chief delegate.) Under terms of the agreement reached, the consortium would stand the cost of the Globe's construction (minus 26 million crowns), even though possession of that edifice would go over to a municipally controlled joint-stock enterprise, AB Stockholm Globe Arena, which was to pay rent to the city for its use. In return *Hovet* would obtain ownership of the land – valued at 170 million crowns – on which the commercial buildings were to be built. Reconstituted as Globen City KB, it also would be entitled to the rents derived from the resulting office and commercial space, whose permissible area had been adjusted upward to 143,000 square meters.[33] In addition, the muncipality not only provided the consortium with an interest-free loan of 375 million crowns,[34] but also invested 180 million crowns in road and other infrastructure improvements around the Globe. County tax funds were also siphoned off to cover expenses for a radical renovation of the subway station most closely serving the complex. Actions ensued one another with such speed, and were conducted with such disdain for ordinary procedure and established working principles, that construction was already under way by the time that detailed architectural plans were approved by local administrative authorities on August 28, 1986.[35]

For those who sooner or later became aware of these circumstances, who learned through the mass media at the time or in conjunction with the reportage of later events, who chatted about them with family or acquaintances, there must have been more than a little bit of puzzlement, consternation, dis-ease.[36]
Backroom deals.
Secret negotiations between politicians and self-interested business representatives.
Businessman-politicians who grease their own wheels.
What in the world was going on here?
Such things don't happen here.
Who among them is to be trusted or believed anymore?
Isn't this a democracy where decisions are reached with careful rationality?
After hearing from all sides?
Isn't this Sweden?
What has the world come to?
Where in the world am I, are we?
And – crumbling-meaning afterthought in the context of the February 1986 assassination of Prime Minister Olof Palme and countless lesser loss-of-recognition episodes –
What does it now mean to be a Swede?
Who in the world am I, are we?

The perpetrators of the agreement facilitating construction of the Globe and Globen City were more than a little pleased with themselves, more than a little content with the thought of what was to result from three billion crowns of production costs, with the thought of 6,000 new jobs by 1992. Some of the politicians involved saw themselves as having performed an act of magic, an "indian rope trick"[37] – not only had their arena been obtained with "a minimum of municipal funds,"[38] but urban land value and a tax base had been created where none before existed. The financial backers of the project had cause for delight – not only had they been able to make another move in a game of "rampant land speculation," not only had they found another office-space outlet for their huge sums of overaccumulated capital, but in so doing they had also gotten around "their lack of control over buildable land, and the lock on project planning and development enjoyed by the municipal government."[39] Now, politicians as well as private-sector enthusiasts exclaimed, Stockholm could compete with other northern European centers for revenue-generating sports and entertainment spectacles.[40] Now, with a "world class" building, Stockholm's position as a "world class" city was confirmed, its niche in the urban hierarchy of the global capitalist economy reaffirmed. Now Stockholm could more than hold its own glob(e)ally, now it could be further internationalized. Now Stockholm had an "Eiffel Tower" of its own, or at least a *Pyramide du Louvre* of its own[41] – a new "neon sign" visible to Sweden and the world at large,[42] a new "collective symbol,"[43] a symbol that would become inescapably imprinted upon the national consciousness with the issuance of four postage stamps depicting it. Now the image and the aura of the Globe would reassert that Stockholm – and Sweden as a whole – was worthy of foreign tourism, worthy of being one of the world's leading convention cities,[44] worthy of major investments, worthy of remaining an important node in international circuits of capital.[45] Now it could be reemphasized that Stockholm was a physically attractive city, with attractive shopping alternatives and exciting entertainments. Now, by virtue of newly materialized symbolic capital, the commercial and industrial sectors of Stockholm – and all of Sweden – would fully realize their developmental possibilities. Now, in addition, a fading industrial area was to be revitalized, while a greater developmental balance was to be struck between the area to the north of the central city – where the Moderate (Conservative) and Liberal parties were predominant and where the commercial construction investments of recent years were disproportionately concentrated – and the southern part of the Stockholm region – a traditional Social Democratic bastion that was short on office employment. Now, in a manner quite different than that suggested by the Stockholm Exhibition of 1930, the interests of Swedish capital and the Social Democrats were once again to be jointly served in a spectacular space.

But would they really?

In the end?

GLOB(E)AL ARCHITECTURE IN THE "POSTMODERN" ERA

EVA GÖRANSSON
(1989)

The Globe is a symbol of Power, impressive and colossal.
　It is size in particular which is of great significance to Power. A monument should be seen, be dazzling, be something to see up to. Yes, size itself should teach us people our place. We should feel our littleness and powerlessness.
　Power should be unimpregnable.... Therefore the materials of Power are also hard and eternal. Stone, bronze and marble. Steel, tempered glass, concrete and aluminum.

[T]he "Tax [Sky]scraper"[46] would fit into the huge sphere. . . . 3,500 [metric tons] of steel had been used for construction beneath "the equator." If one were to lay out all the rods used in the reinforced concrete they would extend from the Globe to the Eiffel Tower. . . . In the roof there are 144 airplane-type windows. Around "the equator" itself there are 100 windows, which are 120 centimeters [about four feet] in diameter.

Dagens Nyheter, February 17, 1989 and ULLA KARLSTRÖM (1989), 16

All the roofing elements were [made] . . . by baking a sandwich of insulation material between two aluminum sheets.

The arena, through its shape and size, will itself become a symbol of the complex [as a whole], widely visible over the city and internally the city's, the country's, Europe's most exciting space – magnificent and yet intimate – for all the grand events we Stockholmers currently miss.

Statement circulated with initial designs for the Globe and Globen City by Berg Arkitektkontor AB, as quoted in Sahlin-Andersson (1989), 106

Those firms which establish themselves here will pick up some of the radiance from what happens within the middle of the ball.

1985 advertising brochure for commercial space in Globen City as quoted in Torbjörn Andersson et al. (1991), 3

The architects were inspired by early modernism to create an urban district of clean geometric forms, where the public arena, built for games and festivities, would dominate the area as a sphere. . . . The twelve-storey Portal House . . . associates among other things to the style of the 1930 Stockholm Exhibition.

HÅKAN ÖSTLUNDH (1989), 92, 96

Whereas the modernists [of architecture and planning] see space as something to be shaped for social purposes and therefore always subservient to the construction of a social project, the postmodernists see space as something independent and autonomous, to be shaped according to the aesthetic aims and principles which have nothing necessarily to do with any overarching social objective, save perhaps, the achievement of timeless and "disinterested beauty as an objective in itself."

DAVID HARVEY (1989b), 66, 85

[Most] postmodernists simply make gestures towards historical legitimacy by extensive and often eclectic quotation of past styles. . . . If, as Taylor puts it, history can be seen "as an endless reserve of equal events," then architects and urban designers can feel free to quote them in any kind of order they wish.

O.K.
WERCKMEISTER
(1991), 74

Now architects and architectural critics charge modern architecture with the vital problems of mass society, in a reversal of their predecessors' claims that architecture could solve them. However, the postmodern critique of the social reform programs which Le Corbusier once summed up in his slogan "Architecture or Revolution" foregoes any political alternative. The new architectural forms it propagates are no longer sustained by the ethos of housing, but by the aesthetics of the built public sphere. At a time when comic-strip artists are picturing fantastic, futuristic metropolises, devoid of decoration, as future sites of oppression and decline, postmodern architects are offering a richly decorated scene, not for work or living, but for the enjoyment of culture.

The Globe, in all of its immensity, was not to stand on its own architecturally. Built at a breakneck pace (2.5 years from initial site work to formal opening for the 1989 World Hockey Championships despite repeated resort to newly improvised techniques), its design was to be integrated with the remainder of the new Johanneshov complex, with the buildings and grounds of Globen (the Globe) City. The entire project was to be physically and thematically unified. It was, in a sense, to appear as a single theme park, one where the spectacular consumption offerings of the Globe were to be the main draw, and the three-tiered shopping arcade passing through several of the main-profit-source office buildings was to be a secondary draw (if for no other reason than its scale, being twice the size of "the Gallery" (*Gallerian*), the largest arcade in downtown Stockholm). In arriving at its winning design, Berg Arkitektkontor AB satisfied these objectives through falling back upon the full register of postmodern architectural devices – pastiche, irony, eclecticism, the appropriation of "local history" and "collective memory," the re-presentation of local architectural icons, the more general "citing" of historical forms and (near schizophrenic) double coding.[47]

Code one. The postmodern as modernism *sans* the social engineering rhetoric of Le Corbusier and the Swedish functionalists, as more modern than modernism in its use of pure and simple lines, as a play upon that spectacular consumption space most pivotal to "local" and national history – the Stockholm Exhibition of 1930 – as spectacular play citing other architectural modernisms expressed during the 1920s and 1930s, as playful (and therefore anti-high-modern) modernism. The Globe itself, among many other things, a much magnified echo of Gunnar Asplund's Planetarium at the 1930 exhibition. The forty flags immediately outside the Globe, fluttering with their abstract predominantly red patterns, a restatement of the wings-

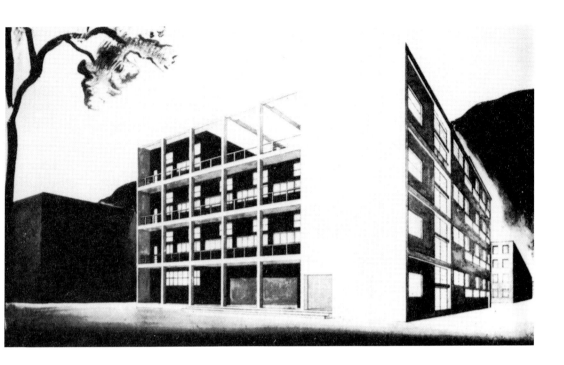

of-progress flags ubiquitous to the 1930 exhibition. A restatement underlined by the ample use of a variety of flags along the approach to the Portal House as well as outside the Stockholm Globe Hotel. Atop the Portal House itself, a steel framework jutting skyward, a constructivist mast engirdled with two red discs, an on-high and yet belittling tongue-in-cheek reworking of that principal 1930 icon, the so-called "advertising mast." Among the several other rephrasings of the 1930 Stockholm Exhibition: the free-standing vents for the 1,200-car garage indistinguishable from some of the exhibition's kiosks; the ship-deck windows of the Portal House quoting some of the *"Funkis"* housing displayed at the exhibition; the large, spare and unmistakably functionalist lettering employed on one of the hotel's side walls to advertise that building's eating establishments. The occurrence of no other colors but bright reds, blues and yellows inside the Globe, as well as the use of nothing but primary colors on the exteriors and interiors of all of the area's other geometrical structures, an intentional quoting of the 1920s Dutch de Stijl school.[48] And then – what is this? – a little architecture-is-not-politics joke, writ very large in the form of the front façade of the luxurious Stockholm Globe Hotel, in the form of a ground-to-roof white grill projected in front of a canary-yellow wall. Here a superimposition of lattice, color and smaller (window-) squares that is not only reminiscent of a Mondrian painting, but is also a direct citing of Giuseppe Terragni's (in)famous *Casa del Fascio* in Como – a "house" for the local headquarters of the Fascist party, one in a long string of "Rationalist" attempts at the "reification of fascism in the form of progressive architecture,"[49] a building whose design rested "on the Mussolinian concept that fascism is a glass house into which all can look: . . . no encumbrance, no barrier, no obstacle between the political leaders and the people."[50] Here an impishly designed joke with a possible double entendre, with a possible oblique reference to the *Corso* of the 1930 Exhibition, with a smothered tee-hee-hee reminder that certain modernist styles can "serve just about any ideological interest."[51]

Code one. Ironic postmodern architectural forms ironically, unintentionally, evoking critical questions.

Why the multiple citing of the 1930 Stockholm Exhibition in a space where the consumption alternatives offered are totally removed from any "progressive" social engineering objectives?

Why the multiple citing of that exhibition at a time (the mid-1980s) when the "People's Home" idea-logics it promoted were in the process of becoming abandoned, were being subjected to an increasingly widespread rejection, were being assaulted with extreme vehemence by the very same speculative and financial interests whose representatives were promoting the *Johanneshov* project itself?

Why the reactionary reference to Fascist form, why the perverse plundering of history, why the recitation of an authoritarian imagery in a space devoted to the arousal of consumer desire, to the supposed crowning glory of "free market forces" set free – individual free choice, individual freedom (to consume)?

Why, in playing at paying homage to "local history" and "collective memory," in playing at giving voice to the popular, is there a complete muffling of the working-class-oriented commercial traditions of the adjacent area,[52] is there little more than the playing of visual practical jokes, little more than the telling of inside historical jokes whose punch lines are situated far outside popular frames of reference?

Was Italian Fascism really nothing other than just another quotable set of events, no better or worse than any other quotable set of events?

In the end, is all this anything-goes playfulness and cuteness, like the architectural playfulness and cuteness of so many other grandly scaled postmodern projects, nothing more than slick and deceptive superficiality – the use of ambiguous surface design to obscure hegemonic designs, to aesthetically pacify with powerful conflict-denying symbolic façades while simultaneously hiding (attempts at) a new joint consolidation of economic and political power?

Code two. A free-standing, spatially distinct entity, an integrated urban unit whose "image ... [is modelled on] the confined sense of a medieval town in which buildings closely surround a cathedral (here, of course, the Globe), and where no open [green] space occurs."[53] The relatively narrow foot-traffic bridge from the subway station to the gravel ridge, or esker, on which the complex sits is a "drawbridge." A "moat" is constituted by the street recessed beneath the "drawbridge," as well as by other surrounding streets and a freeway to the east. At the end of the "drawbridge" a "gate to the city" is framed by the twin 60-meter structures of the Portal House – each of them "medieval towers" – and the enclosed and suspended walkway connecting them. An "impenetrable town wall" is formed by the long bank of rectangular office buildings flanking each of the Portal House towers. The space between the office buildings on one side, and the Globe and the hotel on the other side, corresponds to the town "market-place." And the Globe itself, "like the holy cathedral can be seen for miles around and dominates all life in the surrounding area."[54]

Code two. The *Johanneshov* area as a medieval citadel devoted to the culture of consumption. A spectacular, richly decorated scene, a set of consciously theatrical effects, which – like other built forms belonging to what Werckmeister refers to as "citadel culture" – "is capriciously designed" without "any direct political reference to the present times."[55] An elaborately contrived scene-space devoted to creating a sense of "transitory participatory pleasure,"[56] to the provision of yet another image-laden spectacle in what Debord characterizes as the "society of the spectacle" – the society of all-pervasive commodity relations and images. A grandiosely contrived scene-space thereby devoted, presumably, to a further promotion of "modern passivity," to a further camouflaging of the "secrets" of capitalism's ruling powers, to a reinforcement of hegemonic domination, to a further occlusion of opposition.[57]

Presumably.

But could this double coding, this doubly dosed architectural anesthetic possibly work in full, possibly put all those who encountered it back to sleep?

Could the entire array of materialized messages and meanings embodied in the Globe and the buildings surrounding it succeed in being totally manipulative?

Could those messages and meanings be blankly received without arousing some symbolic discontent, without provoking some cultural reworking?

GLOB(E)AL METAPHORS, GLOB(E)AL METONYMS AND GLOB(E)AL IDENTITY ISSUES

LIZ BONDI
(1993), 86

Both [Marx and Freud] insisted that identity – our sense of ourselves as individuals and as social beings – is constructed through structural processes rather than being innate or pre-given. In so doing, both also implied that there are no necessarily universal or unchanging attributes of human identity, but that differentiation and movement between identities is characteristic of modern societies.

STUART HALL
(1991), 22

[W]hat one always sees when one examines or opens up ... [a national identity is that it] represents itself as perfectly natural: born an Englishman [or a Swede], always will be, condensed, homogeneous, arbitrary. What is the point of an identity if it isn't one thing? That is why we keep hoping that identities will come our way because the rest of the world is so confusing: everything else is turning, but identities ought to be some stable points of reference which were like that in the past, are now and ever shall be, still points in a turning world.... [Even though] Englishness [or Swedishness] never was and never possibly could be that.

Today's Swedish Social Democrats are very reminiscent of the bourgeois political block[58] during the 1930s, and vice versa. Today it is the Conservative and Liberal parties who broadcast themselves as the "people of the future," while the Social Democrats employ former right-wing arguments: Keep an eye on, stand up for [that which is]! Preserve and protect [that which is]! Now it is the Social Democrats who appeal to national feelings — "Sweden is unique"[59] — while the business community in its campaigns spreads posters where national identity is seen as an antiquated remnant of the brave new world. The European Community — the sole vision in which the world is still young — became the property of the bourgeois political block. Eventually the Social Democrats crept to the cross: "We also want to be included." Even culturally the bourgeoisie has moved its position forward and Sweden is currently experiencing its second-over-compensatory culture. Euphoric over the current situation, the bourgeoisie is making up for all the lost years. . . . The triumph of the bourgeoisie resulted in the Globe, a building which only impresses through its size and in which one must search for culture in the traditional meaning of that term.

JAN OLOF NILSSON (1991), 95

The left has become super-Swedish and nationalistic. Now it is the right which is revolutionary and wants to demolish the old. Stig Malm [the leader of the Swedish Confederation of Trade Unions] called the [new bourgeois-block] government's policies "unSwedish" no less than five times. . . . The traditional bearers of nationalism, that is to say the right, are developing into "good Europeans." Simultaneously, European Community resisters within the left . . . claim that Sweden has never been a part of Europe, that we must preserve Swedish individuality.

PETER ENGLUND, historian, as quoted in *Dagens Nyheter*, May 23, 1992

Sweden lost its economic sovereignty when our economy was integrated into a transnational economy. That process has been going on throughout the entire twentieth century, although with extra force since the Second World War. During the eighties the process was accelerated through high business profits which for the most part were invested abroad. In 1990 Swedish foreign investments amounted to 84 billion crowns, of which 60 were invested in European Community countries.

KRISTINA PERSSON and KNUT REXED, economists at TCO, a national union for white-collar workers (1992)

[T]here is a strong connection between the internationalization of the economy and the loss and crisis of senses of community.

ROGER ANDERSSON (1991), 2, 29

In the aftermath of the political and economic changes of the 1980s (a successive adjustment of Swedish politics to the mainstream of European politics, acknowledged in 1990 when Sweden decided to apply for membership in the European Community), as well as the radical changes in Eastern Europe ... it is easy to register a growing tendency of crisis for the collective identity of Swedes. Who are we?

ANITA SJÖBLOM
(1991)

The image of Sweden as a forerunner for other countries was dealt a blow last year. Foreign media reported of the Swedish model being brought to lameness by governmental crisis, [the necessity of] an austerity program and unemployment. It's a lucky thing that we have tennis players and Greta Garbo [who, in death, grabbed the international spotlight once more]....

Sweden [was seen as] having been hit by a system-wide crisis; yes, even a national identity crisis.... The Swedes believed they had solved all the world's problems and thought themselves subject to but one difficulty – a shortage of domestic crayfish.[60]

Title of an autobiographical political-position book by CARL BILDT, leader of the Moderate (Conservative) Party, released during the 1991 election campaign which resulted in his becoming Prime Minister. Halland is a province on Sweden's west coast

Hallander, Swede, European.

ORVAR LÖFGREN
(1992)

Do you have a feeling for Swedish form? You appreciate the quality and care of German design. You admire the Italians for their elegance and boldness.

But your heart belongs to Swedish design. You are yourself a part of the tradition which has brought forth Swedish designers. You have a feeling for color and form, know how to combine function and decoration.

You feel it. And one can tell by looking at you.

That's what an advertisement for EGO eyeglass frames sounds like in 1992. Its playing on "the Swedish" is typical of the spirit of the times prevailing in the early 1990s.

ANNA CHRISTENSEN, Professor of Civil Law, in a column appearing in *Dagens Nyheter*, July 27, 1992

There has been much talk about "the only road" lately. In Sweden, today, it is the market economy which is the only way. But there have been, and continue to be, many roads which in their time and place have been regarded as "the only road." Marxism was long "the only road" in a large part of the world. Religious fundamentalism or nationalism are "the only road" in other parts. Common to all of these "only roads" is the conviction that all *social questions can be solved through [the application of] a single, overriding basic principle....*

The market economy seems to be on its way to becoming as fundamentalistic {as the Marxism of the former Soviet bloc}. The basic principle of the market economy will steer everything from weapons production to health care and environmental protection.

A world's record will be set during the days of the World Hockey Championship.... [T]he world's largest illuminated advertisement to date will be projected on the exterior of the Globe with the help of colored lights and letters which will be twenty meters high. According to the Guinness Book of Records, the current record dates from the late 1920's, when a car dealer used one side of the Eiffel Tower for a large illuminated advertisement.

Dagens Nyheter, April 19, 1989

Nobody is likely to have escaped the fact that Sweden was endowed with a new temple this past weekend. Monumental buildings such as the Lund Cathedral, the Uppsala Cathedral and the Stockholm City Hall can feel themselves degraded. From now onward the national sanctuary is called the Globe – but sanctified in the name of what? ... Like all temples throughout the entire history of the world, the Globe should be seen primarily as an expression for contemporary ideas and the spirit of the times.

"Georgie" (1989)

When one approaches Globen City from the slaughterhouse district the colossal office building towers up like a mighty flagship with fluttering banners. The sight is truly impressive. But is it a sinking ship we see?

Dagens Nyheter, September 25, 1992

Whatever the intended messages, whatever the intended meanings, whatever the (would-be) hegemonic designs of the architects responsible for the Globe and Globen City, there were a number of other ways in which the exteriors and interiors of those buildings could be symbolically read during their political emergence, construction and eventual operation. The Globe, in particular, lends itself to metaphoric and metonymic images, being a crystallization of – and contributor to – the ongoing processes through which Sweden's commodity society is becoming further globalized and further subject to unchecked market rule, through which economic and political relations binding the local and the non-local are ceaselessly modified, through which the practices of everyday life, and their associated power relations, are consequently ceaselessly reconfigured. These processes in which the Globe is enmeshed are frequently synonymous with people experiencing a disjunction between the old and the new, with people experiencing an incompatibility of old and new meanings, with the unmooring of once secure meanings, with problems of re-cognition, uncertainty and the no longer reliable, with the undermining of collective memory and taken-for-granted differences (and samenesses). These processes are thereby also frequently synonymous with the widespread appearance of personal and collective identity problems, with identity politics, with struggles over the preservation or redefinition of identity.[61] Or the metaphoric and metonymic images evoked by the Globe frequently link up with the identity issues, with the complexities of (multiple) identity construction, with the identity "crises" which have pervaded Sweden during the late 1980s and 1990s. Time and time again, the images – and related discourses – evoked by the Globe directly or indirectly link up with a pair of fundamental questions:

Where in the world am I, are we?
Who in the world am I, are we?

Through the conjunction of these metaphors and metonyms, through the questions they repeatedly lead to, the Globe may be regarded as a hypersymbol of Swedish hypermodernity, as the (fragmented) totality of late 1980s and early 1990s Sweden in one big Stockholm nutshell.

It was made clear from the very outset by those associated with AB Stockholm Globe Arena (the operators of the Globe) that the entertainment offerings to be made available to the public would be "guided and self-regulated" by "free market forces."[62] Here, despite municipal involvement, there would be no effort to establish priorities or to balance different program types. Here, according to Ingemar Josefsson, priority would be given to the agents and arrangers who offered to pay the most.[63]

The Globe as a place that at one and the same time is showplace and market-place. The Globe as a market-place for the global entertainment industry; as a market-place for the "youth" and "high" music culture, for the spectacular sport culture, commodified by multinational capitalist enterprises.[64] The Globe as a showplace for commodity fetishism in the extreme, as a showplace for marketing people as consumption goods, as purchasable things, as desirable objects. The Globe as a showplace for marketing well-performing products – the show-biz wealth of (the U.S.A., the U.K. and not so many other) nations. The Globe as a showplace for marketing stars and superstars, for marketing commodified bodies of national and world renown.... Global "youth culture" and "light entertainment" commodities marketed in place at the Glob(e)al market-place: Holiday on Ice, Disney on Ice and the Count Basie Orchestra; Elton John, Diana Ross and Bob Dylan; Paul Simon, Cliff Richard and Tina Turner; Prince, Fleetwood Mac and Earth,

Wind & Fire; Frank Sinatra, Depeche Mode and Deep Purple; Rod Stewart, New Kids on the Block and MC Hammer; Guns 'n' Roses, Liza Minnelli and Alice Cooper; Julio Iglesias, Janet Jackson and Simple Minds; Cher, Bruce Springsteen and Ray Charles; Sting, Shirley MacLaine and so on. Global "high culture" commodities gone "light entertainment" commodities marketed in place at the Globe: Placido Domingo, Luciano Pavarotti and a performance of *Aida*. Global spectator-sport commodities marketed at the Globe: Jim Courier, Boris Becker and Ivan Lendl; Sergei Bubka and Carl Lewis as well as the tennis player Stefan Edberg, the high-jumper Patrik Sjöberg and others of Swedish origin. And, of course, the Globe as a market-place for that Swedish "youth culture" commodity become global "youth culture" commodity: Roxette.... The Globe as market-place where commodified bodies are used to market other commodities, where the jerseys and pants of ice-hockey players are covered with advertisements for global firms that market cars and household electronics goods in Sweden (Opel, Pioneer), with advertisements for Swedish firms that globally market steel, cars, trucks, food products and insurance, with the advertisements of Swedish retail chains and coffee firms with global sources.[65] The Globe as showplace where athletic performers become little more than sandwich-board men – than modern counterparts to those plodding downtrodden men who were once a common sight on downtown streets – and at the same time become somewhat akin to prostitutes, being commodities and sellers at once.[66]

The Globe as a confirmation of the global power of capital, as physical confirmation of the global "triumph" of free market forces. The Globe as an urban-landscape reaffirmation of capitalist power, as a built structure completed in 1989, the year in which the economic, social and political structures of Eastern Europe and the Soviet Union were either totally collapsing, or rapidly giving way. The Globe as a locus for free market forces at play, as a profit-making establishment. The Globe as a node in a transnational network of entertainment-commodity circulation (AB Stockholm Globe Arena is a member of the European Arena Association, a co-ordinating organization whose other members include arena operators in Barcelona, London, Milan, Munich, Oslo, Paris and Vienna). The Globe as a facility for facilitating local and international capital accumulation.... Banners and signs prominently displayed in the Globe's various spaces promote foreign-based corporate giants operating in Sweden (IBM, SONY), Swedish-based corporate giants operating in numerous foreign countries (Volvo, Electrolux), and a host of other Swedish companies having either some significant overseas sales, some foreign employment, or some dependence upon physical and financial inputs from abroad. When leased, each of the forty-two luxury corporate boxes yield annual rents equivalent to $50–60,000 (exclusive of ticket charges). In each of its first three years of operation (1989–91) the revenues generated by well over one million paying customers were reportedly more than sufficient to offset event-related expenses, including the annual rent of roughly $2.0 million paid to the municipality of Stockholm.[67] Yet further income was derived from the usage of the building for annual stockholders meetings and for the introduction of new Volvo models.

The Globe as physical manifestation of globalized capital accumulation, of free market forces in action. The Globe as concretization of interdependence patterns within the global capitalist economy, as material testimony to the global sourcing of local construction (and manufacturing) projects.... Suspended near the physical center of the Globe is a four-screened "Jumbotron," a giant television monitor – 4.2 × 3.2 meters – supplied for replay and advertising purposes by the Japanese electronics giant, SONY. (At the time of its installation, it was the largest such [re]viewing device in the world outside of Japan.) The upper hemisphere framework of the Globe was acquired from MERO-Raumstruktur GmbH, a German steel construction company which out-competed bidders from Finland and Japan. MERO, in turn, secured the aluminum façade panels from the Swiss-based Alu-Suisse. Arena seats were procured from another German firm, Heinrich Kamphöhler GmbH & Co. Cooling tower fittings were purchased from Baltimore Aircoil International.[68]

The Globe as a distilled expression of free market forces operating under new hypermodern circumstances, operating in the shifting multiplicity of guises that have become the most salient characteristic of globalized capital accumulation over the past two decades. The Globe as a palpable indicator that capital accumulation no longer rests on Fordism, no longer rests on rigid "long-term and large-scale fixed capital investments in mass-production systems that precluded much flexibility of design and presumed stable growth *in invariant consumer markets*." The Globe as physical proof that the "rigidities of Fordism" no longer reign, that "flexible accumulation" is what primarily makes the world of capitalism(s) go around and around. The Globe as hard evidence that the "motion of capital" has become "more flexible," that capital accumulation has come to rest increasingly "on flexibility with respect to labor processes, labor markets, products, and *patterns of consumption*," has come to reemphasize "the new, the fleeting, the ephemeral, the fugitive, and the contingent in modern life."[69] ... The interior of the Globe is highly flexible, is designed to accommodate "as many different activities as possible,"[70] is capable of being reconfigured to serve a variety of purposes, to serve a variety of entertainment-consumption markets, to meet the particular distinction-reinforcing demands of different groups, to satisfy audiences with different tastes. The Globe can be adjusted to hold anywhere from 3,000 to 16,000 spectators through the use of retractable stands at its lower levels, of removable chairs at its floor level, of a spotlight framework that can be automatically hoisted up and down and fixed at different heights, of dismountable stages, of drapery and stage wings and of ice-covering rugs. Within the space of a few hours it may be converted from any one of the following configurations to any one of the others: ice-hockey or ice-show arena; rock-music concert hall; boxing stadium; meeting or convention hall; basketball arena; choral or opera concert hall; tennis stadium; premiere-evening movie theater; circus ring; track-and-field arena.[71] The physical flexibility and architectural design of the Globe are such that, regardless of configuration, the entire attending (consumer-)audience has a sense of being close up, right in the middle of "the action," in intimate proximity to the performing commodity it has purchased. And, in order to reinforce this sense, "when the Globe is not full the empty seating can be drawn back so as to give the impression all the seats are taken."[72]

The same "free-market," "market-force" discourses which accompanied the further globalization of Stockholm during the Reagan–Bush and Thatcher years, which echoed the ideology of the American and British right wing (with a heavy Swedish accent), which provided a rationale for the deal enabling construction of the Globe, and which were bound up with so many of the metaphors and metonmyic images evoked by the Globe, were also tightly interwoven with the discourse surrounding the most dramatic change confronting *fin-de-siècle* Sweden – the possibility of full membership in the European Community as well as any eventual politically united European Union.[73] Those Conservative Party leaders, other politicians, business-community figures, academics and editorial writers who projected the "neoliberal" market-force discourse asserted, among other things, that Sweden's economic future hung in the balance, that a choice lay between the competition-saves-everything Heaven of membership and the economic-isolation Hell of non-membership.[74] Without membership, it was contended, Swedish producers of goods and services would be left out of the world's largest free-trade market. With membership, however, Swedish firms would be able to compete on an equal footing with the domestically based firms of Germany, France, the United Kingdom and other member countries. And, because answering to only one set of rules and standards, Swedish enterprises would thereby be able to avoid the expenses of separately adjusting to domestic circumstances, to achieve greater economies of scale and scope, to have significantly better chances for long-term expansion – not least of all because of a greater facility to compete globally with U.S. and Japanese corporations. With membership, Swedish firms would be put on their toes, would be made subject to increased competition within the country's boundaries, would – in the aggregate with their new competitors – offer a greater variety of alternatives, would be driven to provide better products at lower prices, all of which would result in expanded individual "freedom," in a greater freedom of choice for consumers, in consumers becoming the "big winners."[75] Moreover, with membership Sweden would remain a viable investment alternative for internationally mobile investment capital of both domestic and continental origins; whereas without membership Swedish corporations would continue to prioritize offshore investments – especially within the protective walls of the commuity from which it was excluded – while foreign investors would for the most part continue to regard the country as too geographically peripheral to be of any interest.[76] This in turn, according to a fear-inspiring image promoted by the Federation of Swedish Industries *(Industriförbundet)* might well result in lower wage levels (in order to keep Sweden somewhat attractive), diminishing prosperity, and further social-welfare reductions (owing to a shrinking of tax revenues).[77]

In the face of such power-ful arguments, which turned primarily on the promised Paradise of increased consumption, on an implied renewal of the "Great Consumption Party,"[78] and in the face of additional supportive arguments which insisted on Sweden's ability to exert significant influence on the full range of policies developed in the Community's (Union's) various political institutions, a series of counterarguments appeared and a heated debate emerged on the wisdom and consequences of Swedish membership. (The pro-membership arguments issued from the country's economic and political power bases may themselves be seen as a defensive strategy developed in anticipation of counterhegemonic discourses.) Around workplace coffee-break tables, in casual household and social-occasion discussions, as well as in guest newspaper columns and television talk shows, these (freely paraphrased) were among the questions being daily raised by dissident and unconvinced voices during 1991 and 1992[79] – voices often rhetorically dismissed and demeaned as belonging to "provincialists, folk-dance fanatics and women who don't understand economics and want everything to remain as it is."[80]

If any product that has been approved in one member country must be automatically approved in all other countries, if the standards for electrical appliances and other goods sold in Sweden are not to be determined in Stockholm, what is to happen to old guarantees of "Swedish quality" as well as to the quality of Swedish life?

If an enormous bureaucracy in Brussels is to have jurisdiction over the details of everyday life in Sweden, if Sweden must loosen up its environmental controls and terminate its state-run monopolization of alcoholic beverage sales, what is to remain of Swedish sovereignty?[81]

What is one to make of the principle of "subsidiarity" – the principle of decentralizing political power to the lowest possible level of government, of allowing "decisions to be made as close to the citizens as possible"?[82]

Does it mean that most actions will be taken either in national parliaments or provincial-level governments, or does it really mean that power will be concentrated in the hands of those who determine what is "as close to the citizens as possible"?[83]

What remains of our self-determination, of our ability to keep joblessness to a minimum, if the called for monetary union results in a German(-like) central bank which vigorously prioritizes low inflation levels over low unemployment levels?[84]

Does not all of this mean a transfer of power to distant, faceless and uncontrolled bureaucrats who are well beyond the earshot of public opinion?

Does not all of this mean the end of (national) parliamentary democracy as we know it?

The end of Sweden as we know it?[85]

Implicitly or explicitly, these and other questions led to another series of questions and realizations regarding one's (feared loss of) identity as a Swede, one's (future or possibly already diffusely existing) identity as an European,[86] one's (renewed, security-reinforcing, closer-to-home, materially grounded, face-to-face rooted, sense-of-place related) identity as a resident of a particular region or province.[87] Or, in the wake of the "free-market," "market-force" discourse associated with both the proposed Europeanization/globalization of Sweden and the Glob(e)alization of Stockholm, in the wake of disintegrating meanings and an increasingly unrecognizable present, and with prospects of a totally unfamiliar future confronting them, people were in essence asking themselves:

What in in the world is going on here?
Where in the world am I, are we?
Who in the world am I, are we?[88]

The Globe as an embodiment of class relations and class segregation, as an objectification of the fact that those who have greatly succeeded in the "free market" are frequently above it all, frequently beyond the reach of those below, frequently beyond the reach of normal legal restrictions.... The Globe's luxury corporate boxes, like their counterparts in North American stadia and arenas, are situated in the highest tiers of the facility. However, it is not only from those privileged perches that corporate executives, financial wheeler-dealers and their guests may peer down on the performers and people beneath them, may enjoy their consumption of entertainment in social and spatial isolation. On one side of the arena, a "luxury" restaurant physically belonging to the adjacent Stockholm Globe Hotel juts out at the "equator," just above the seating area. There, behind a floor-to-ceiling window and beneath nineteenth-century crystal chandeliers, the well-to-do put "first-class raw materials" prepared with "a large portion of creativity" into their mouths,[89] put expensive "world-class" gourmet dinners under their belts while putting the evening's "world-class" commodified event under their gaze. There, from their shielded on-high position, from their superior situation, from their removed roost, members of the expense-account set drink in the evening's contest or concert with their eyes while drinking away at their exorbitantly priced pre-dinner cocktails, dinner-time aquavit shots or wines and post-dinner cognacs – all the time being legally set off from the common world by a glass wall, being technically exempt from the Swedish law which forbids the consumption of alcohol while attending sporting events. In that gilded cage, in that protected upper space of the upper classes, in that feasting nest far removed from the ice-cream, pizza and hamburger bars of the foyers, the common letter of the law is not a text for the uncommon.[90]

The Globe as a world unto itself, as a physical distillation of the space-compressing, space-blurring world of contemporary commodity capitalism(s). The Globe as a physical distillation of the world in which "free market" agents find a way of commodifying just about everything, find a way of colonizing just about every niche of everyday and everynight life with some form of consumption – and in the process not only repeatedly compress the local and the global into one another, but also repeatedly dissolve the distinction between public and private spaces. (This occurs, among other ways, through the privatization of public urban space, through the conversion of public civil space into private market space, and through the creation of privately operated spaces, such as the Stockholm Globe Hotel's Arena Restaurant, which are designed at one and the same time to entice people to shop or otherwise consume in public and to convey the ambiance of dwelling-space intimacy.)[91]

... In the Globe, a *public* arena in which embodied entertainment commodities are offered for sale, the paying audience is eased into plush red seats of easy chair-like comfort and confronted with the "Jumbotron," the television set of all television sets. Here, then, the spectacle consumer is supposed to feel as if she is in an environment not unlike the secluded haven of her own living room, as if in that private space where the public world makes its daily flittering-image incursions. Here, right-in-your-face advertisements also punctuate the flow of on-screen entertainment, thus inverting the public invasions occurring at home and making them larger than life.[92] Here, many experience the public, for-private-profit, spectacle in a manner much like they experience spectacles in the privacy of their own domestic space. After being here, a young man may observe: "The Globe is a building for people whose interest in sports was founded in front of the TV. There is no sense of solidarity or community at the Globe. To sit there is like sitting in front of the TV"[93] (except that the Globe's giant screen allows some possibility of one becoming a spectator to one's own spectatorship). Here, in this liminal space where the public and the private are transformed back and forth into one another, where attendees are within roles associated with both, and yet betwixt the two, where attendees are likewise suspended between culture and economy,[94] television technology is employed to erase spatial distinctions in another more insidious manner. At the entrance to the Globe, and at one location within the building itself, as well as at a number of consumption sites elsewhere within Globen City, television cameras (and security guards) are used to keep an ever vigilant eye on people, to help maintain order and protect property, to discourage "inappropriate behavior," to discipline and contain spontaneity, to discourage the congregation of youths and the scrawling of graffiti, to assert uninterruptedly that seemingly public spaces are under private control.[95] Here, in a Sweden where free market proponents call for an end to the bureaucratic surveillance exercised by state institutions, free-market agents themselves exercise physical surveillance via a persistent electronic gaze. Here, if "Big Brother" is not to watch you, "Big Capital" certainly will (with the help of the police who maintain the outdoor cameras).

The blurred transposition of private and public occurring in and around the Globe, as well as the relentless further commodification of everyday life, was given another expression through hyperenergetic efforts on the part of neoliberal politicians to privatize or otherwise radically restructure welfare and other public services following their election in September 1991.

Those setting party policy and most of those making municipality-level decisions knew all the answers in advance of any specific question, knew full well how to incant their action-justifying neoliberal mantra. The unrestrained operation of market forces will best satisfy all individual and societal needs, will cure all social ills, will remove all the difficulties and social dissatisfaction created by Social-Democratic planning, The free reign of market rules will inevitably make things better. The practices of the Social Democrats, the self-styled "Party with the Good Gifts," are passé. The public sector can no longer be expanded every time a new problem appears. On the contrary, the public sector is incapable of solving problems because it is inefficient, overly bureaucratic, characterized by "low productivity," and operates outside the reach of competition. The time has come for a "change of systems." The era of the "guardian state" is over. Statism is dead. The end of high taxes and the state's role as impersonal guardian will result in greater economic incentives, in greater personal autonomy and a liberation of the imagination, in greater individual freedom for people to do as they want, to spend as they want, to consume more.

The municipal privatization and cutting back of daytime child-care services, the reduction of emergency-room and other health services, the elimination of after-school programs, the scaling-down and localized privatization of home care for the elderly, and other similar measures were legitimated not only by such oft-repeated rhetoric, but also by another, sometimes overlapping discourse regarding "civil society." In the new civil society, the impersonal care provided by state-controlled, "totalitarian"- (Soviet-) like, tax-financed institutions would be replaced by the efforts of people bound up in "good networks," by family, friends and relatives, by neighborhood or church groups and other small-scale forms of social community, whose members would unconditionally and lovingly take care of one another without pay. The state's "forced solidarity" institutions would be no more. Christian ethics would reassert themselves in the informal institutions of "the small world." (A fictive, never existent) Paradise would be regained. A new rule of thumb would prevail: "If something needs to be done and it can be done by civil society, then it shall be done by civil society, and not by the state or municipal government."[96]

Before long, many personally affected people from across the political spectrum were left emotionally reeling, feeling bewilderment, profound discontent, or extreme dis-ease.[97] They lapsed into a puzzled or dissatisfied state as welfare institutions were swiftly put either entirely on their own or semi-privatized (contracted out to firms that were given subsidies or direct public payment for services performed); as at least one public library was privatized and countless others had their allocations slashed; as the public monies paid out to theaters, museum and cultural institutions were drastically reduced; as all these measures were given urgency by the reduction of state contributions to already suffering municipal coffers and the worsening fiscal crisis of the state itself (resulting from, among other things, an already high debt, the recent introduction of lower income tax rates, artificially high interest rates, surging unemployment and the more general impacts of a global recession). As changes ensued one another with unanticipated speed, shocks of disjunction, of uncertainty, of lost meaning and new language became entangled with crises of personal confidence and identity (which on occasion were mobilized into grass-roots protest actions).

What will become of me if my nursing, library, social-service or municipal office job is terminated, reduced to part-time or put under the control of a private firm?[98]

What is to be done about the elimination or paring back of care services for our children and aged parents?[99]

Won't most of the burden fall on me and other women in similar circumstances?

Doesn't "market forces" plus the "new civil society" add up to large numbers of women losing their public-sector employment, to already overburdened working women and newly unemployed women being forced to take up the welfare and personal-care slack?[100]

Doesn't the civil-society rule of thumb really mean: "If something needs to be done and it can be done by unpaid female labor, then it shall be done by unpaid female labor, and not by women paid by the state or municipal government?"[101]

Won't matters get even worse for women if Sweden joins the European Community and our social policy comes under the influence of Germany?

How are we to believe that market forces will eventually solve our welfare problems if the unregulated involvement of the nation's banks in massive real-estate speculation and shady deals has contributed both to depression-like economic conditions and the fiscal crisis of the state?

If we really do want the welfare state to be further dismantled, if we really do want the state to have less power, how are we to counteract the injustices and inequities constantly generated by the market system, to cope with its inevitable production of losers as well as winners?

What are we to make of the new meanings and terms of those in power?

"Condensen" means to eliminate child-care or other service facilities?[102]

A "positive savings strategy which yields development possibilities" means a budget reduction?[103]

And, what does it now mean to be a Social Democrat if the Social Democratic Party is heavily implicated in the present set of circumstances?

If the Social Democratic leadership took a number of steps throughout the 1980s which in effect committed them to "market-force" solutions?[104]

If they were the ones who initiated the reduction in state payments to the municipalities?

If they at the same time cut taxes in a manner that seemingly benefits the wealthy?

If they no longer appear unequivocally committed either to social equality or to full employment (that which has long been on the top of their agenda)?

If they are at least in part willing to go along with the eventual elimination of about 150,000 public-sectors jobs through "rationalization" (efficiency measures) and privatization?

What in the world is going on here?

Where in the world am I, are we?

Who in the world am I, are we?

The Globe as a microcosm of yet another national buying binge, of an extended shopping spree on the part of Swedish consumers in general, of the so-called "Great Consumption Party." The Globe as a distorted, fun-house mirror image of the increased levels of household consumption that came after another Social Democratic concession to "free-market forces" – the further deregulation of the credit market that occurred in November 1985, when the Central Bank of Sweden (*Riksbanken*) was deprived of its administrative capacity to restrict bank loans to households and private persons. The Globe as an all-encompassing consumption sphere, as a commodity-celebration space, as a space in which to consume commodified spectacles from all over the world while being adjacent to a subsidiary mall space in which commodified goods from all the world can be purchased for consumption. The Globe as THE site for holding the "Great Consmption Party" – a several-year party in the course of which Swedes almost obsessively sought to give meaning to self and life via the (over)consumption of goods and services, via running up the industrialized world's largest household debt, via spending more than they had at their disposal.[105] The Globe as THE site for a several-year party in the course of which repeated episodes of immediate gratification and intoxicated giddiness were further enabled by a simultaneous decline in oil prices and a fall-off in the value of the dollar (which reduced the cost of many imported goods). The Globe as THE site for a several-year party whose characteristic over self-indulgence was doomed to result in a national hangover, was doomed to help bring about an eventual decline in domestic manufacturing through the increased purchase of foreign products, through investment-discouraging high interest rates brought on by an unfavorable balance of payments, and through decreased savings and thereby a decreased supply of investment capital.[106] The Globe, with its own simple geometric form, as a condensation of Swedish "tasteful reservation" on a spending rampage, as a paradoxical reflection of a continued commitment to the streamlined "Swedish modern" designs of the 1930s[107] amidst widespread dissatisfaction with the social engineering/solidarity ideology originally associated with those designs.

The Globe as both a spectacular cathedral and a cathedral for spectacles. The Globe as "the church of the new era, an homage to global power, where light comes from above, and where the powers [that be] are again found at the very top."[108] The Globe – complete with awe-inspiring dome and window-filtered skylight suggestive of the sacred and the mystically communal – as a spectacular worshipping place for the newly popular cult of the ungoverned market economy, for those devout believers in the scriptures of "neo[economic] liberalism." The Globe as a building appropriate to a period in which religious status has been conferred upon a doctrine proclaiming that unfettered "free-market forces" are the solution to all social problems. The Globe as a place at which to bear witness, to have uplifting experiences, to achieve inner peace via the consumption of spectacles, via the exposure to performances. The Globe as a cathedral in which community is to be attained, in which everyday estrangement is to be erased by spectacle, in which transubstantiation is to occur, in which the commodity is to be transformed from an instrument of alienation to a substance of togetherness.[109] The Globe as a cathedral space in which the entertainment commodity and religion become indistinguishable from one another, in which the sacred spectacle and the profance spectacle are placed on equal footing, in which "high" and "low" spectacular forms are reduced to the same level of exchange – to just another "experience" to be given the "big sell." . . . One of the first widely distributed advertisements for "Globen City – the office area with its own sports arena" read in part as follows: "In February the Globe will be inaugurated. The first big event will be the World Hockey Championships in April. In June Pope John Paul II will hold a catholic mass. In October its time for the Stockholm Open [an international tennis tournament]."[110] . . . Pilgrims to the Globe paid hefty ticket prices that in some instances well exceeded the equivalent of $200 when Verdi's Requiem was given a spectacular rendition in the spring of 1992, when Pavarotti took on the principal solo and 1,800 singers were rounded up for a "World Festival Choir" from throughout Sweden as well as Denmark, Norway, Finland and the U.S.A. Moreover, when the unwary approached the cathedral, they were hooked into paying almost $20 for a "special program," even though a free program awaited them on their seats.[111]

The Globe as a temple for worshipping celebrities, for paying reverence (and hard earned income) to embodied commodities, for paying devotion to entertainment- and sports-world deities. The Globe as a temple not only for worshipping those who – regardless of national origin – have distinguished themselves in contemporary global competition. The Globe as a temple for also worshipping those Swedes of the long and recent past who made it to the top in "high" or "popular" culture markets. . . . The extensive foyers of the Globe are here and there decorated with large photographic portraits of "Names we never forget," with black-and-white (and thereby "truly" historic) pictures of the dead and the no longer active who gave Sweden an international name, with personal images that contribute to some binding together of a highly individuated population. These are profiles that are meant to evoke shared memories, to reawaken collective pride in bygone record-book performances, heroic feats, team victories and artistic accomplishments, and thereby to renew a sense of national identity. These are faces that fit into a rewriting of twentieth-century Swedish history in keeping with the (would-be) hegemony of free-market interests. In short, a pictoral rewriting of history which focuses on individual (commodified) success, which focuses on people who at once represent both the essence of "Swedishness" and involvement in the world beyond Sweden's borders – and, yet again, relegates domestic social conflict to the obscure, if not the invisible. A gallery of Swedish goods that sold well on foreign markets, that made good internationally, that made Swedes feel good about themselves, that reaffirm a minimum cultural consensus:

> the "Grenoli" trio – Gunnar Gren, Gunnar Nordahl, Nisse Liedholm – who brought great success to Sweden's national soccer team in the late 1940s and 1950s, who were the first Swedes to succeed as professionals abroad, as commodified players in Italy;
> Greta Garbo, the mysterious money-making machine, Sweden's all-time gift to Hollywood;
> Björn Borg, the twice thingified, who attached his name to a line of apparel when he tired of serving as an attraction at Wimbledon and elsewhere;
> Jussi Björling and Birgit Nilsson, whose exceptionally rich voices yielded exceptional riches through performing on opera and concert stages throughout the world;
> ABBA, the pop-song quartet who learned to spin platinum from the flimsiest of straws;
> Ingemar Johansson, former "Great White Hope," former world heavyweight boxing champion, former maker of millions in a matter of minutes;
> Astrid Lindgren, who translated her name into millions by creating "Pippi Long-Stocking" and other storybook figures that proved translatable into a long list of major and minor languages;
> and so on;
> and so on.

As neoliberal and other (would-be) hegemonic interests moved Sweden toward European Community membership, efforts were made to rewrite Swedish history so as to (re)instruct the population at large as to where in the world they were, as to who in the world they were. Highly characteristic of these efforts was a film about Sweden and its history that was produced for the Swedish pavilion at the Seville World's Fair and shown with some advance publicity on national television during the summer of 1992. A thoroughly hypermoden, rapid-fire sequence of fragmented images intended to make people "feel Swedish," a flickering sequence of images which emphasized the rituals and traditions associated with key dates on the annual calendar (*Valborgsmässoafton* [Walpurgis Night], Midsummer's Eve, Sankta [Saint] Lucia), the deeply sedimented character of certain institutions and the seasonally shifting beauty of the landscape. A thoroughly hypermodern stream of quickly passing images which at the same time emphasized that Sweden was once a great power on the Continent, that it has not always been neutral, that it has a history of being engaged in Europe. While an ambitious, definitive sounding series of museum exhibitons – "The Swedish History" – was being prepared for 1993 with the intent of "shaping a national identity which is common to all,"[112] one revelation after another was made which suggested that neutrality was a fiction, that Sweden actually never had been neutral.[113] This latter array of messages was in keeping with a larger set of widespread representations: we have always been a part of Europe, through our trade, through the experience of our journeymen, through the long past in-migration of Walloons and other groups, through our sharing of the culture of Shakespeare, Molière and Goethe, through our sharing of modernity and modernism, through what we have taught in our schools, through and through and through.[114]

The interior of the Globe as a dreamscape where pleasure is derived from being a spectator, "just looking," visually consuming the product of the day. The interior of the Globe as a phantasmagoriascape whose fleeting images and sensations are confined to the entertaining, as an escapescape where everyday concerns may be at least momentarily forgotten, as a distractionscape where daily drudgery or disappointment may be put out of mind by staged excitements, as a wish-fulfilmentscape where promises of gratification may be fulfilled – until the reverie is no more, until re-entering the troubled outside. The interior of the Globe as a diversionscape where personal woes and societal crises may be temporarily ignored or repressed through identifying with the performers, through thereby reaffirming some dimension of one's own (collective) identity, through the consumption of cultural commodities packaged in spectacular lighting and sound effects. The interior of the Globe as an up-in-the-cloudscape where one may yell and cheer, where one may transgress the emotional-control boundaries of "civilized" everyday life.[115] The interior of the Globe as an illusionscape, a world where all that exists is worry-free and easy to understand, more or less predictable and yet exhilarating. The interior of the Globe as a paradoxical fantasyscape, a never-never-land where growing unemployment, business failures, banks on the border of bankruptcy, welfare-system cutbacks and other current problems may be briefly whisked from consciousness through the operation of "market forces," through the purchase of products which appear most profitable to the arena's operators. (Those cultural products which appear most profitable to the Stockholm Globe Arena Corporation often appeal to people in their late teens and twenties, to that element of the population perhaps most hard hit by current job-market contraction and economic uncertainties.) The interior of the Globe, lined with blood-red seating, as a womb-like setting, as a warm, comfortable, nurturing world of problem-free consumption, closed off from the local, national and wider world of economic instability and social conflict.

The Globe as a commodified escape space which – also being a nodal space in the global circulation of sports and entertainment commodities – cannot always escape the world of politics, the world of struggle and turmoil beyond Sweden.... In February 1991, at the height of the Persian Gulf War, local police authorities deemed it necessary to sharpen security at a major international track meet held at the Globe so as to prevent any "acts of terrorism" being directed at Carl Lewis or any of the other U.S. participants.[116] ... Seven months later, the Globe's management was forced, on fairly short notice, to cancel a concert scheduled for Ray Charles owing to the fact that he remained on a UN blacklist of artists who had performed in South Africa. (Despite the repeated application of pressure, Charles had refused to apologize for his 1981 appearance.[117] By the spring of 1992, when events in South Africa had turned somewhat, the Globe's bookers found it possible to arrange a July concert.) ... In keeping with the active pro-ANC stance of the still-governing Social Democrats, the Globe had been given over, in March 1990, to a "gala evening" to celebrate the recent release of Nelson Mandela.[118] There, accompanied by Prime Minister Ingvar Carlsson, Mandela gave his first public address outside of South Africa since leaving prison. The speech itself was sandwiched in spectacle: Miriam Makeba and numerous Swedish pop-artists singing before and after its presentation; the ANC logo flashing in laser light; the obligatory all-stars-on-stage chorus at the end.

If Sweden were actually to gain admission to the European Community, if free-market zealots and their otherwise motivated supporters were to have their way, then the country's government had to fully accede to the Maastricht Treaty. In order to do so, the Swedish state eventually would have to participate in "a common foreign and security policy, including the ultimate creation of a common defense policy which with time can lead to a common defense."[119] The required advance acceptance of an as yet unformulated and unagreed upon common defense policy was plainly incompatible with the parliamentary resolution backing Sweden's formal application for European Community membership, as it specified that the country's "policy of neutrality" was to be retained. In the summer of 1992, when the European Community Commission released its "notice" regarding the suitability of Swedish membership, and thereby identified those issues which would be most sensitive in upcoming negotiations,[120] the problematic nature of any continued neutrality was underscored and the already ongoing public debate over membership versus neutrality grew ever more intense and further spilled over into the informal discourses occurring within the spaces of everyday life.[121]

Neutrality is not merely a policy that Sweden has pursued in one way or another since the mid-nineteenth century, but a value-laden term highly charged with taken-for-granted meanings, a term long regarded as unforfeitable, a term deeply sedimented in the collective consciousness of much of the Swedish population. "Neutrality" has served as a symbol for the nation's freedom, for its position outside of power-bloc alignments, for its consequent ability to take an independent stance on important international issues. Its various meanings have contributed to a widely held self-image of Sweden as a bastion of foreign-policy altruism, as a guardian of international morality, as a voice of sanity in a world of senseless conflict fed by vulgar patriotism and extreme nationalism, as a political defender of otherwise trampled upon Third-World peoples. But now the meaning(fulness) and necessity of neutrality had been thrown into question. Now that sometimes pride-filled term (which is congruent with perhaps the most widespread of Swedish personality traits – the avoidance of conflict, the internalization of any comment or reaction that might hint of aggression)[122] would have to be heard and seen differently, if not totally surrendered to hegemonic claims.

Prime Minister Carl Bildt: "A policy of neutrality was self-evident during the decades when Europe was divided into blocs.... But today [with the end of the Cold War, with the Warsaw Pact no more, several former communist states expressing a desire for NATO membership, and no overhanging nuclear threats] the situation is fundamentally changed.... [I]nstead we see how unsolved national, economic or ecological problems lead to one conflict after the other.... It ought to be obvious that it is only through deepening cooperation in all areas that we can meet the broad spectrum of peace threats in the new Europe … which we are a so self-evident a part of."[123]

Sverker Åström, Social Democrat and former Undersecretary of State for Foreign Affairs: "[The debate] ought to deal less with the obvious incompatibility of neutrality and membership and more with the possible alternatives to a policy of neutrality that no longer serves its purposes."[124]

Anita Gradin, Social Democrat, Ambassador to Austria and former Minister of European Affairs: "As a UN member Sweden has already [demonstrated] a preparedness to relinquish its neutrality."[125]

Sven-Olof Lodin of the Federation of Swedish Industries: "There is no international role today for a Sweden that stands outside. What shall we be neutral towards? Towards peace? Towards stability? Between which forces should we balance when we sit on the seesaw?"[126]

While a number of politicians began dropping "neutrality" from their vocabulary, replacing it with "non-aligned" (or "alliance-free") and insisting that membership did not require involvement in NATO or any other military alliance,[127] fear-filled questions something like the following were repeatedly raised in private and public.[128]

How can we be certain that future circumstances in Russia or elsewhere won't require our neutrality?
How will we be able to stay out of any war that might develop in Europe?
Won't the abandonment of our neutrality mean that Swedish soldiers could be forced to serve in other countries without our approval?[129]
Won't the termination of our neutrality open the possibility of nuclear weapons being stored on our soil?
Or might it not allow the remainder of the European Community to force us into introducing trade restrictions against a war-engaged nonmember?

Both before and after the membership-versus-neutrality debate heated up, variously motivated statements and releases to the mass media, as well as a new frankness in Russia, helped to demythologize, or at least call into doubt, Sweden's past neutrality. What was "Medelsvensson," the "Average Swede," to make of his country's neutrality during World War II given renewed accounts of German troops being carried to Norway via the Swedish State Railways, of crucial iron-ore and munitions shipments to Germany, of the massive letter-opening action conducted by the security police against actual and suspected anti-Nazis at least until the battle of Stalingrad?[130] How were people to construe their country's neutrality during the Cold War given new evidence that a Swedish plane shot down by the Soviets in 1953 was gathering radio-signal information, some of which was to be forwarded to U.S. intelligence agencies; that far-reaching co-operation had long existed between Swedish defense planners and NATO; that the Swedish military knew full well that the much publicized intrusion of Soviet submarines into Swedish waters was very likely motivated by knowledge that West German U-boats with offensive capabilities, as well as advanced U.S. listening devices, were being permitted to operate in those very same waters?[131]
With the persistence of a sometimes bewildering debate, with revelation heaped upon revelation, with a newly documented past catching up with and melting into an unclear and multiruptured present, with a growing recognition that representatives of the state – the supposedly reliable, rational, democratic

state – had been lying to the public on neutrality-related issues, with a new awareness of having been intentionally deceived year after year, decade after decade, with the meaning of neutrality being redefined and reredefined into collapse, the path toward Europeanization/globalization once again led countless women and men to pose a set of identity-wrenching questions.

>   Can this really be Sweden?
>   What in the world is going on here?
>   Where in the world am I, are we?
>   Who in the world am I, are we?

The uncomfortable urgency of these questions was reinforced by another domestic intrusion of the world of politics beyond Sweden, by another set of ongoing events that was incompatible with the widely held self-image of Sweden as an international moral force, as a moral superpower, as a model of altruism, as an exceptional island of racial and cultural tolerance, as a refuge for the politically oppressed, the ethnically or religiously persecuted and the war-torn. As the former Yugoslavia disintegrated in the early 1990s, as first Croatia and then Bosnia-Herzogovina were ravaged by war, as the barbarity of "ethnic cleansing" spread, as the Serbs further restricted the rights of ethnic Albanians dwelling in the province of Kosovo, Sweden was beseigned by refugees from those areas at the same time that asylum-seekers and refugees continued to arrive from other parts of the world. (During 1992 alone about 40,000 Kosovo-Albanians were admitted to Swedish refugee-processing camps. Over the same twelve months more than 28,000 people had arrived from other parts of the former Yugoslavia in hope of being given permission to remain in Sweden, pushing the total number of asylum applicants in the country up to around 83,000).[132] Confronted by an unprecedented budget deficit in the vicinity of 100 billion crowns, by negative economic growth and rapidly mounting unemployment, the newly elected "bourgeois" coalition undertook a succession of inconsistent measures to stem the refugee flow while issuing a stream of statements which were frequently contradictory or self-serving. Many could well wonder: what in the world is going on here? First the Conservative-led government overturned the Social Democrats' tightening up of refugee policy (thereby indicating that Sweden was once again open to those who did not fit the Geneva Convention's narrow definition of a "refugee").[133] Then they used various rationales to send back Somalis and Iraqi Kurds arriving via Russia and Estonia, to reject Jews who claimed to be severely persecuted in Russia (suggesting that they instead go to Israel), to turn away desperate Bosnians for want of a visa agreement with their newly proclaimed state, and to attempt the deportation of draft-avoiding, discriminated and oppressed Kosovo-Albanians, thereby ignoring the advice of the UN High Commission for Refugees.[134]

If scarce resources were being pressed by the arrival of 15,000 new refugees per month,[135] the (inter)nationalist Swede could certainly understand the Minister of Immigration Affairs, Birgit Friggebo, and the Minister for Social Affairs, Bengt Westerberg, when they jointly declared: "Sweden cannot alone, or together with a few other states, take responsibility for the worst European refugee situation in fifty years. In such a situation Sweden could be forced to take measures ... which we would preferably avoid for principled and humanitarian reasons."[136] The (inter)nationalist Swede could also easily appreciate statements by Prime Minister Bildt and others to the effect that Europe as a whole "ought to fulfil its duty to stand up for freedom and human dignity,"[137] that other countries such as Denmark, Finland, France and the United Kingdom ought pull their fair share of the refugee load.[138] She or he might even accept claims that Sweden could best serve Bosnians and other ex-Yugoslavians by contributing to the cost of camps in Croatia and other "neighboring territories" where those refugees could remain until the cessation of hostilities would permit them to return home[139] (even though such claims might appear naïvely far-fetched in light of copious mass-media evidence regarding the property appropriations associated with "ethnic cleansing" operations). But, who in the world was one if the Prime Minister could suggest that every time Sweden accepted Bosnian Muslims it was further encouraging the Serbian "ethnic cleansing" program; if the Minister of Immigration Affairs saw nothing "offensive," "objectionable" or "shocking" in the turning away of Bosnians; if she, the director of the National Immigration Authority, and the Prime Minister on various occasions publicly complained of the Kosovo-Albanians as "traditional petty thieves," "criminally inclined" and "increasingly involved in criminality" – despite contradictory testimony provided by the police chief for one of the municipalities where they were encamped?[140] And who – in the midst of this ever-more unrecognizable world – was one if Georg Andersson, the Social Democratic Minister of Immigration Affairs from 1986 to 1989, was correct in making the following proclamation?

> The question is whether a humane refugee policy has ever existed. It is true that for many years Sweden has received relatively many refugees.... But the sacrifice has been slight. Our bragging about generosity and humanity has not been in reasonable proportion to our deeds. It is also true that over the years we have sent away a large number of asylum applicants. Those permitted to stay have seldom been treated as equal fellow beings. It has been difficult for them to get work. They have been excluded from the activities of our associa-

tions and social influence. They have seldom been [socially] admitted to our community.[141]

Who – in an ever more confusingly globalized Sweden – was one now if one had come around, like the majority of one's fellow citizens, to favoring a reduction in the number of Muslim refugees admitted to the country; if one had come around to regarding all male Muslims as fundamentalists who deny any rights to women;[142] if one more readily, yet involuntarily, laughed at jokes, degrading rumors and apocryphal anecdotes regarding "primitive," "stupid" and "naïve" refugees or immigrants.[143] Who in the world was one now if at the same time one felt uncomfortable or incensed about frequent encounters with big lettered BSS (Keep Sweden Swedish) graffiti, about the scattered violent actions of skinheads and members of VAM (White Aryan Resistance), about the occasional firebombings of homes and stores owned by Third-World immigrants, about the desecration of Jewish graves in the Stockholm area, about the serial shooting of non-whites by the notorious "Laserman," about the racist rhetoric of the populist New Democracy Party, about the flowering of racist and anti-semitic publications, about resistance here and there to the establishment of a local camp for refugees already in the country, about the growing numbers of people who evidence an "atavistic impulse directed against all things alien?"[144]

THE GLOBE(AL) THEATER: OR, ALL THE WORLD IS A STAGE AND THE PEOPLE ON IT *NOT* MERELY PLAYERS RECITING HEGEMONIC LINES

[T]he notion of globalization as a noncontradictory, uncontested space in which everything is fully within the keeping of institutions [of capital], so that they perfectly know where it is going, I simply do not believe.

STUART HALL
(1991), 32

For cultural analysis and criticism, the contesting of the meaning of things or events is what centrally constitutes politics.

GEORGE E. MARCUS
and MICHAEL M.J.
FISCHER
(1986), 153

Even as "the market" makes its profits, it supplies some of the materials for alternative or oppositional symbolic work. This is the remarkable, unstable and ever unfolding contradiction of capitalism supplying materials for its own critique.

PAUL WILLIS
(1990), 139

DAVID HARVEY (1993), 23	[I]maginations are not easily tamed to specific political-economic purposes. People can and do define monuments in ways that relate to their own experience and tradition.

An informant (male, civil servant, 25 years old), EMSU	*The Globe is an unnecessary show-off [or bragging-] building.... The best thing about the Globe is that it is round. Its best side is from the outside. A piece of graffiti raised to the second power would be to paint big black pentagons on the Globe so that it would look like a soccer ball [that presumably could be kicked away. Out of sight!].*
Letter to the editor, *Örebro-Kuriren*, April 1, 1989	*I was almost born on a soccer field and like all sports. But if one is as poor as a louse, one has to be content with seeing top-level sports by long distance.... The admission prices to sports arenas like the Globe are purely fantastic.... Now capital has taken over sports and we who are born on the shady side don't have a chance to see top-level events.*
An informant (male, professional consultant, 34 years old), EMSU	*A friend of mine has told me that one morning one of his friends who lives on Söder [Södermalm]*[145] *went out on the balcony and looked at the soon completed Globe. Then he turned inward towards the apartment and said to his live-in girl-friend: "Look what a gigantic piece of dandruff is lying out there."*
PER SUNDGREN, Left Party member of the Stockholm Municipal Council as quoted in *Dagens Nyheter*, December 22, 1990	*It's a complete joke to claim that the Globe is only costing the taxpayers 26 million [crowns].*
KAJSA O., columnist, in *Dagens Nyheter*, January 18, 1992	*Of course, in the beginning I was enchanted by the Globe when it popped up in the path of the proposed extension of Goth Street [Götgatan]. But I pretty quickly tired at the sight of it. And now, when one has to pay the bill for something other people have made money on, one begins to look at the latest Stockholm monument with other eyes.... Blowtorch a notch in the building and use it as a piggy bank.*
INGEMAR UNGE (1992), 70	*[Through a nearby living-room window] the Globe really obtrudes like a gigantic eye, sitting in some space-monster who's got into his head to peek in at Mrs Bölke.*

Graffiti art in a Stockholm passageway during the spring of 1992: "The Globe is watching you!"

Although appealing to many, the Globe is too big physically, too much of a spectacular space, too frequently the site of well-publicized spectacles, too frequently encountered in the course of everyday life, too much of an omnipresence, too much of an ever-present eye in the sky, too thoroughly enmeshed in "market-force" discourses, too implicated in the Europeanization and further globalization of the city, too tainted in its historical origins, too much the consequence of a collaboration between political and economic powers, too much a theater of the present, too closely identified with issues touching upon identity, too much a concretization of meanings to go culturally uncontested, to escape symbolic reworking by discontented Stockholmers and other Swedes. From the time of its construction onward, this site of commodified entertainment, of commodified bodies providing commodified culture, has proven a site of cultural politics. The Globe as concrete form and concretization of meanings has not always been greeted passively or given an enthusiastic reception. It has proven a site where the aesthetics of power have been subjected to symbolic discontent, where architecture and politics have been given a gleeful popular twist, where physical form and (would-be) hegemonic message meanings have been challenged from below, where an enormous object of consumption has been creatively reworked so as to confirm position and identity.[146] It has proven a site where meaning-filled monument and monumental new meanings have been simultaneously called into question openly or in the hidden interstices of daily life, where folk humor and word warfare have repeatedly come into play, where those who are supposedly ideological easy prey have frequently not been easily preyed upon, but have instead constructed their own experience- and (op)position-based statements.

The very naming of the Globe itself was a counter-counterhegemonic stratagem, a move on the part of the powerful to ward off the symbolic night-raids of the locally displeased, of those who refuse to leave the excesses of power verbally unassaulted, of those notorious for their counterhegemonic deployment of folk humor, of those feared for their quick resort to the stinging word, to the weapon of the weak long most preferred locally. Originally the Court [*Hovet*] Consortium responsible for construction and development had settled upon the name *Glob*, the indefinite form of Globe in Swedish,[147] but quickly changed their minds in light of the word's unsavory physical associations in English.[148] This stratagem was not entirely successful, as mention or sight of the Globe, via a chain of associations, might still evoke the indefinite *glob*, might still be connected to bodily fluids. Asphalters whose street work brought frequent visual exposure to the Globe at some point began referring to it as "the Spitball" (*Spottloskan*).[149] Among Stockholm University students who were sometimes concert or ice-hockey match attendees it became common to say: "Shall we go to the Spitball tonight?"[150] And, within the offices of the Swedish Union of Commercial Employees [*Handelstjänstemannaförbundet*], there are some who spoke of the Globe as the "Snotball" [*Snorkråkan*].[151]

Until construction of the Globe was completed, Göteborg, the country's second largest city, was one up on Stockholm in terms of spectacular symbolic capital because of the Scandinavium, a gracefully shaped multipurpose arena capable of attracting entertainment and sports events that otherwise would have gone to the capital. With the Scandinavium at least partially eclipsed, with the image of Göteborg as the future hub of all of Scandinavia dealt yet another blow,[152] with yet another confirmation that the city would always remain secondary to Stockholm, local wits used the Globe as a new focus of attention for their anti-capital-city barbs. It served as a new target that at one and the same time enabled them to reassert their local identity, to vent their frustrations, and to rework their sense of perpetual economic, cultural and personal(?) inferiority. Tired of all the national-media publicity surrounding the opening of the Globe, some inhabitants of Göteborg took to reducing that enormous solid-skinned sphere to a thin-membraned bubble, to an air-filled nothing, to an overblown joke, referring to it alternatively as "the Show-Off (or Bragging) Bubble" (*Skrytbubblan*),[153] or "the Comic Strip Speech Bubble" (*Pratbubblan* – the bubble in which the words of comic figures are enclosed).[154] In a commentary in Sweden's largest daily newspaper, a prominent Göteborg author, Anderz Harning, wrote with venomous irony of his recent visit to "the Venice of the North," a "little ice-floe-surrounded city ... hit by a new religion," a city where "the newly saved tell of the Birth," a city where "The people worshipped a little ball. A glob(e). It glistened a little in the sullen fall sun, but there was nothing more than that to it."[155] In order to cope with what some locals designated "The Globe Syndrome," other Göteborgers commonly word-whipped the hated building by diminishing its form to nothing more than that of a winter vegetable, by speaking of it as the "Cabbageseum," or *Kålleseum* (pronounced much like coliseum).[156] Doubly pun-ishing, this term also evoked images of buffoonery, of *Kålle* – a mythical Göteborg working-class figure who long has served as the butt of the "*Kålle och* [and] *Ada*" jokes told on variety-theater stages and commonly repeated in private spaces. And, as a result of a contest held on a popular local radio program, the Globe was mercilessly converted either into an object already past its prime, "Baldie" (*Kalhuvet*), or into an object that on first reflection could be thought of as the superfluity of an upper-class woman, as "the [billowing, rounded lower half of an] Opulent Ball Gown" (*Festblåsan*).[157] *Festblåsan*, moreover, verbally cut in two other mocking ways. It could simultaneously be seen as something bloated with urine, the "Party (or Festival) Bladder," and heard as a (weak-sounding, self-promoting) "Party Horn."

Aside from its position in a wider discursive terrain and its obtrusive daily palpability, the Globe provided Stockholm residents with numerous sources of disgruntlement, dissatisfaction and annoyance, and thereby with numerous reasons for the expression of symbolic discontent. Not only did people become gradually aware of the questionable circumstances under which the Globe's planning and construction was contracted out and financed.[158] Not only did it sooner or later dawn on many that the Globe's development involved a circumscribing of democratic procedures and Social Democratic complicity in the privatizing of Stockholm real estate. Not only were many taken aback by the eventual revelation that the Globe and Globen City were costing local taxpayers far more than they had initially been told (at least 425 million crowns rather than 26 million crowns).[159] Not only did many of those queuing up for scarce apartments grow angry over news that the Globe project was helping to drive construction costs up at least 20 per cent and slowing down the rate of new housing completions.[160] There were also those who were less than pleased with the break between the new consumption and office land uses of the *Johanneshov* area and its former working-class connections with nearby *Söder* and the adjacent slaughterhouse and stockyard district; between the new spectacular spaces and the home stadium of the working-class supported Hammarby soccer team which had to be partially torn down in order to make room for the project.[161] There were those in the residential areas in the near vicinity of the the Globe who grew bitter over newly introduced parking problems, over the local traffic chaos re-occurring with each big entertainment event, over the exorbitant parking-violation fines risked by visiting friends and relatives.[162] There were also those among nearby residents who were driven to protest over the removal of a pond where children used to play among the frogs and small lizards, over the trees that had been felled for ground-clearing purposes, over the legality of the few sorry trees planted on the broad walkway between the Globe and the office buildings of Globen City.[163] There were those from throughout the Stockholm metropolitan complex who were occasionally moved to grumbling anger either over the cost of tickets, over the difficulty of obtaining tickets at all, or over the drinking privileges accorded those who viewed the Globe's spectacles from the restaurant in the Stockholm Globe Hotel. There were the approximately 50,000 incensed people who were turned away in wintry weather at the gates of the Globe after purchasing cheap firm-sponsored tickets for a preview tour of its interior (little more than 100 were admitted before the doors were closed for "security reasons").[164] There was even the occasional patriot who grew hot under the collar because the flags "outside our new pride" were "hard to define [and] red-like," because the blue and yellow national flag was not even flown there on June 6, on the national holiday known as Swedish Flag Day (*Svenska flaggens dag*).[165]

Whatever the source of their discontent, those Stockholmers symbolically reworking their knowledge and experience of the Globe – those constructing their own counter-meanings for that commodified entertainment structure – did so largely through the use of nicknames. Although a discontent-free spontaneousness was sometimes involved, the nicknames that were invented and tossed around in daily parlance, like most other nicknames attached to people and objects, were generally mocking or irreverent, were usually punishing in their intent, and yet all the same often contained an element of ambivalence, of reluctant praise combined with abuse of some kind.[166]

HANS AXEL HOLM
(1992)

[I]f there is no [other] meaning to it, it is at any rate a [puny] ball, a monument to the swindling 1980s.

Almost without exception, nicknames played upon the Globe's spherical form. Perhaps the most innocent labels used were "*Kulan*" and "*Kalaskulan*," translatable most simply as "the Ball" and "the Party Ball."[167] However, as *Kulan* carries a number of connotations, these terms could often be heard rather differently, in ways that communicated irony and were likely to elicit a snigger. If *Kulan* means ball, it usually conveys the image of a ball in miniature – a marble, a necklace bead, a pellet – and thereby impishly suggests that the mighty Globe is really something puny, something of rather little significance. Moreover, as *Kulan* also translates as "the Bullet," "the Cannonball" and "the (track and field) Shot," these nicknames could be taken to mean that the Globe was something dangerous, something hurtling through the air at you, something that could destroy you. A further popular designation, aurally linked with *Kulan*, was *Bulan*, a term which lacked any of the former's hint of aesthetic praise, as it translates as "the Bump," "the Bruise," or "the Swelling," and thereby suggests that the Globe is something painful, something unwanted, a physical malformation, an injury to the urban landscape.[168] Another cluster of nicknames described the Globe as a particular kind of ball – as a white tennis ball (*Den vita tennisbollen*), as a golf ball (*Golfbollen*) (perhaps even "the world's biggest golfball"), as a bowling ball (*Bowlingklotet*), as a ball in the air (*Bollen i luften*).[169] While each of these appellations points to some appreciation of the Globe's shape and alludes to its function as a sports arena, the latter two in particular suggest an element of unfixity, an element of unreliability on the loose, a capacity to land on your head, to knock (bowl) you or something over. Other nicknames combine an apparent appreciation of form with a (conscious or subconscious) sense that the Globe is something foreign, something that doesn't belong where it is, something which is either out of this world, or of another world – "the Great Mosque" (*Den stora moskén*), "the Spaceship" (*Rymdskeppet*) or "the stranded spaceship on the tundra," and "Moonbase Alpha" (*Månbas Alpha*).[170]

Surely, it's Josefsson's testament to posterity.

BENGT BROBERG, chairman of the board, *Djurgården* Sporting Club, as quoted in Randi Mossige-Norheim (1988), 12

Most of the other nicknames pinned on the Globe by Stockholm residents are more like *Bulan*, more unmistakably abusive, more aggressive in their expression of discontent, more carnivalesque in their obscenity, in their reference to the "lower regions" of the body, to bodily wastes and to the frailties of the body's upper half. For some the Globe is seen in much the same way as by those Göteborgers who verbally diminish it. For them it is a not-to-be-taken-seriously "comic strip speech bubble" (*Pratbubblan*), or perhaps a "bald head" (*Flintskalle*), an object whose sun-reflected gleam is no better than that of a shiny pate.[171] A few who "were completely against construction of the Globe and thought the money could have been used for something more sensible" took to calling it "the world's largest scrotum-ball" (*pungkula*), a single huge testicle, impressive in its size but – implicitly – not worth much in its splendid isolation.[172] But the most politically and dis-affection-laden nickname does not merely begin and end with a giggling sexual reference, with an earthy view of the Globe, with a ribald rounding attack on that building's roundness. Instead it cuts with a double edge. Instead it also verbally degrades a detested power holder. It lewdly lowers a specific politician. It fiendishly ridicules Ingemar Josefsson – he who was the prime mover behind the creation of the Globe, he who used his seat on the municipal council to push through the free-land-for-construction-cost deal, he who earlier had sat on the financial-plan and design competition jury, he who made light of what the actual costs to the muncipality might be, he who was the quintessence of the Social Democrat turned market-force advocate, he who contended that the best cultural spectacles for the "city's largest party-place" were the spectacles whose promoters paid most, he who managed to retire from the municipal council at age 52 with a "golden parachute," he who eventually became Chairman of the Board of Directors of Stockholm Globe Arena Incorporated (AB), he, who in the eyes of his detractors, wished to attain immortality by "creating something that nobody could avoid seeing."[173] For some among the working-class grassroots of the Social Democratic Party, the Globe became (and remains) "Josefsson's glans penis," or "Josefsson's penis head" (*Josefsson's ollon*).[174] Here the skyward-pointing curvature of the Globe is not converted to anything as socially proper as a bald head, but to a dick head. And not just any dick head. But the dick head of one who is regarded as a dick head. What a way to culturally contest, to rework a rounded form into a phallic symbol, into a representation of phallocentric political (and economic) power.[175] What a way to recognize that the (urban and other) planning schemes of politicians are themselves phallic symbols "whereby the present penetrates the future," are themselves "efficient technique[s] for raping the future," for "preserving what now is by transforming fleeting intentions into [the] unyielding stones of physical [and institutional] structures"[176] which subsequent generations can often only lie back and unwillingly accept. What a way, at the same time, to signify the disillusionment and dissatisfaction felt throughout Sweden among those at the grassroots bottom of the Social Democratic hierarchy toward those no-longer-in-contact, out-of-reach power holders self-isolated at the top.

The symbolic reworking of the Globe was not confined to everyday verbal expressions, to folk-humor forays, to nicknames that subversively slighted and maliciously mocked. In a special comic strip that ran on Sundays in the country's largest newspaper during the summer of 1991, the Globe was the central representational element of a story-line which spoke to the not always submerged collective (and contradictory) fantasies of much of the Swedish population (if the political opinion polls of the time are to be believed).... An "insidious, sly, treacherous company of individuals," who were motivated in considerable measure by "a blind desire for profits and power-hungry," had plotted to blast the Globe off its moorings so that it could roll northward down the length of Goth Street (*Götgatan*), across the islet containing the

THE GLOB(E)ALIZATION OF STOCKHOLM 229

scenic Old Town and onward until it crushed the Swedish Parliament building.[177] In the political chaos supposedly to follow, the wily schemers would push through new housing legislation that would permit construction of "the Cube," a new residential structure occupying almost all of the Old Town, measuring one kilometer on each of its sides and containing eight million apartments, or enough units to provide individual housing for the country's entire population. Once the Globe is actually unmoored and in motion, the heroes of the story dynamite the Tax Administration Building (*Skattehuset*) so that it obstructs Goth Street, deflecting the mammoth spheroid eastward until it harmlessly lands KA-PLASK in the water.... A number of things are going on in this strip, not all of which are captured in this synopsis, and each of them is subject to multiple readings. Here – among other ways of seeing and hearing the contents – we have the embodiment of spectacular consumption set free, assuming one of its nickname forms (*bowlingklotet*, or the bowling ball), rolling on its merry way to knock over that source of rules that intrude into everyday life and hinder unbridled consumption, to destroy the highest seat of political authority, to bring down a government as yet dominated by the Social Democrats, by those only recently and still often ambiguously committed to the freeing up of market forces. But – newly hegemonic market-force discourses or not – rampant speculation in real-estate development has come to be seen in a bad light, disallowing any fantasy-world triumph by villains who conduct such activity. Instead, the heroic James Dog and his sidekick go one step better than the "bad guys" in clearing the way for more consumption. They fell the "Tax (Sky)scraper" (*skatteskrapan*), that imposing bureaucratic phallus, that concretization of the commanding–demanding–appropriating Father, that obstacle of all obstacles to unrestrained consumption and greater "individual freedom," that site where Stockholmers annually surrender monies which otherwise would be spent on goods and services. They demolish that site fronted by a desolate wind-blown plaza where every February 15th a carnival like atmosphere erupts as women and men leaving in their forms at the last minute dance in the streets, listen to musicians perform, parade political message banners, shout anti-tax slogans in chorus, wave their arms at passing drivers until they toot their car horns, and in other ways ritually appropriate the space of those who appropriate from them, ritually invert the bureaucratic order-liness of the building beside them.[178] Here, in this comic-strip world ambivalence abounds. Here actions against the very same speculative forces that produced the Globe and Globen City are also designed to enable more consumption. And yet those actions result in a praised(–abused), loved(–hated) object for entertainment consumption being done away with, winding up in the water, sunk out of sight. Here is a realization that, with the conditions prevailing in early 1990s Sweden, the discursively pumped-up desire for much lower taxes and much higher consumption can only be satisfied in a fantasy, in a dream that is – in the end – something of a wet dream.

GLOB(E)AL CRISES AND DISASTERS: OR, WHAT GOES AROUND, COMES AROUND

Cocky Locky, Cocky Locky, the sky is falling! It's the end of the world!	HENNY PENNY (freely rendered)
The whole country, indeed the whole world, is in a shaky state.	JOSEF KOVACS, Liberal Party municipal council member in Haninge, a Stockholm suburb, as quoted in *Dagens Nyheter*, September 19, 1992
The Globe has become an exciting addition to the cityscape. It is also a good example of how the business community and the city of Stockholm can collaborate with happy results.	PER PALMGREN, Conservative Party member of the Stockholm Municipal Council as quoted in *Dagens Nyheter*, February 16, 1989
I think that the Globe makes a lively contribution, even if one must be a realist and recognize that it is perhaps only in Stockholm that the Globe is world famous.	LENNART RYDBERG, Liberal Party member of the Stockholm Municipal Council, as quoted in *Dagens Nyheter*, February 16, 1989
The Globe is a "very successful affair" both for the city of Stockholm and the consortium which built the arena and the appendant office complex, believes [Mats] Hulth [Social Democrat and member of the Stockholm Municipal Council] who thinks the Globe is today worth one billion [crowns].	*Dagens Nyheter*, December 22, 1990
After the bourgeois [coalition's] election victory in [September] 1991, the private sector of the economy has been blooming, this report [of the Swedish Trade Council] makes apparent.[179] *The Stockholm correspondents of foreign newspapers, [the report continues], write of how good everything has become since the shift of government.*	*Dagens Nyheter*, June 26, 1992
Here we stand now, in the midst of a hangover from the financial market's champagne binge of the 1980s. Within the space of a few years real-estate	ÅKE LUNDQVIST (1992)

prices multiplied. The money lending of the banks joyfully kept pace. No matter how quickly building prices rose, the banks did not hesitate to give new loans to the new real-estate firms.

Finally the crash came. A crash which is costing taxpayers many billions as a result of the [government's] "crisis package" and subventions to insolvent banks.

And the people curse. Curse the real-estate sharks who left the debts for us, but rescued a personal fortune for themselves written over to their wives and children. Curse the bank directors who were forced to resign but also got millions in severance pay.

LENNART RYDBERG, as quoted in *Dagens Nyheter*, January 22, 1992.

Purely economically the Globe is not such a great deal, that's obvious. The calculations were very shaky right from the start. I fear that this year's loss is only the beginning of big economic troubles.

ROLF WIKLUND, consultant, as quoted in *Svenska Dagbladet*, September 25, 1992

The shopping center [or mall] within Globen City has been marketed all too poorly. When one comes [on the adjacent freeway] one can hardly see where to get off. Moreover, many visitors have thought that the shops were located in the Globe itself and they consequently have been disappointed.

BO G. BOESTAD, Chairman of the Board of Directors for Globen City KB, as quoted in *Dagens Nyheter*, September 25, 1992 and *Svenska Dagbladet*, September 25, 1992

A project like the Globe area always demands capital infusions in order to keep going during the first years and now the owners are saying stop.

The area as such has been a success. The problem has been that it was expensive right from the start. Above everything else, it was the costs for constructing the arena that went wrong.

Dagens Nyheter, September 25, 1992

"This is a pure shock," [says Randi Danielsson, owner and operator of Jam Jam, an ice cream and candy store situated on the mall within the Globen City office complex]. Finally the disappointment is too great for her and she begins to cry.

On September 24, 1992,
in the midst of economic turmoil throughout the network of globally interdependent capitalisms,
in the midst of an extended global recession,
in the midst of difficult economic times in Germany, Japan and the United States,
in the midst of Sweden's worst post-war economic crisis,
in the midst of a prolonged fall in the country's gross national product,[180]

in the midst of unemployment rates soaring to levels unheard of since the depression years immediately following the Stockholm Exhibition of 1930,[181]

in the midst of declining industrial production and a cascade of factory closures and cut-backs,[182]

in the midst of Alice-in-Wonderland interest rates imposed so as to buttress the krona against the German mark and thereby prove that Sweden was earnest in its desire to join the European Community, truly eager to further Europeanize/globalize,[183]

Globen City KB – the consortium that built the Globe and now operated the office, hotel and retailing space of Globen City – ceased making payments, declaring that, with debts already in the vicinity of two billion crowns, its partners no longer wished to infuse additional funds.[184] It was but four days after the Conservative-led government and the Social Democrats had announced their "historical compromise," a "crisis package" of austerity measures to deal with the country's battered economy. It was many months since "crisis" had become a frequently used term in the everyday vocabulary of Stockholmers and Swedes in general.

The dizzying circumstances that had prevailed until 1989 – when the first occupiable space in Globen City became available – were now a matter of the past. The days of but three years ago – when the demand for offices in Stockholm "was higher than ever" – seemed long past.[185] The dumping of "clean" and "dirty" money into huge construction projects was at an end. The injudicious overspeculation of the 1980s had come home to roost. The willingness of banks to make no-questions-asked loans to fly-by-night and shell-game real-estate firms had just about evaporated. Real-estate prices were rapidly tumbling in Stockholm. The costs for "broken pipe-dream calculations" could not be passed on to Globen City's office and shop occupants, who could pack up and leave for vacant and more cheaply priced downtown space at any hint of a rent increase.[186] In fact: "The deep real-estate crisis in Stockholm had forced the management of Globen City to lower their rents."[187] Two-thirds of the space in additional Globen-City buildings that were nearing completion was as yet unrented. None of the space in a third round of construction was as yet leased.[188] Now, loan-providing banks and financial institutions – already hard hit by forfeitures and the collapse of other real-estate projects – were not alone in standing to lose money unless a debt rescheduling agreement could be reached.[189] A portion of the interest-free 375-million-crown loan made to Globen City KB by the city of Stockholm was also as yet unrepaid.

Whether or not the management of Globen City KB was at all justified in (dis)placing the blame for its financial woes on the costs incurred in constructing the Globe, the building and the company which ran it were themselves swept up in the wider economic crisis. Although Stockholm Globe Arena Inc. had just about broken even in 1991 (losing 274,000 crowns, or about $50,000), losses for 1992 were expected to surpass 22 million crowns according to a projection made some months before the economic fall-off of Stockholm and Sweden as a whole had accelerated into a tailspin.[190] Numerous Swedish and foreign-based multinational corporations, themselves suffering from global difficulties, were pulling out of their commitments at the Globe. Half the luxury corporate boxes stood empty after being fully rented the previous three years. Stockholders' meetings and other corporate events were being switched to less spectacular, less costly sites.[191] Advertising revenues were declining. The rent to be paid the city was up 20 per cent and was to be increased again in 1993.[192] The number of concerts and other non-sports events booked was considerably lower than in 1991, and consequently total attendance was down. As income sources became ever-more leaky, severe leaks in the building became all the more evident with each heavy rainfall.

Report from a self-proclaimed "luxury," "world-class" establishment[193]

> What did we encounter [at the restaurant of the Stockholm Globe Hotel]? Two bottom-class lunches, the first time involving half-uncooked fish of the modern anonymous sort [and] uncooked vegetables with rice. All of it cold and insufficient.... The second day we had chicken, another dish which they had made into a big failure. Totally tasteless and insufficient! ... Forks down for restaurants poorly run by amateurs.[194]

Commodified "disaster" spectacle at the Globe as spectacular disaster

In a time and place where the spectacle has become an everyday consumption good, in a time and place where the occurrence of crises and disasters with a global dimension are a part of everyday awareness, why not make a commodified spectacle of disaster? Why not commodify the misfortunes and misery of others?

During late August 1991, the Globe was the site of several performances of "Disaster-91," "the world's largest disaster play," a much ballyhooed spectacle designed to give "society's rescue services" a "unqiue chance to show how they work," a drama enacted by 150 policewomen, policemen, firefighters, nurses, doctors and ambulance drivers.[195] Those paying the price were to be thrillingly entertained by a scenario in which: a (presumably foreign) "terrorist" hijacks a schoolbus; a crash occurs between the bus and a tank truck; a daring rescue of injured children is made; the "terrorist" subsequently retreats into a dairy with a young girl as a living shield; an entire residential area is simultaneously evacuated owing to ammonia spilled from the damaged truck; and a special police squad eventually captures the "terrorist." At the first presentation the Globe was scarcely half filled. The total expected attendance for all performances was 140,000; but the actual number was 72,000. The organizers lost about one million crowns.[196] A spectacular (economic) disaster. Potential attendees had apparently preferred real drama to the commodified simulacrum. That weekend many Swedes sat glued to their television sets, fixed on direct-broadcast spectacular images of the failed coup in Moscow.

The Globe's spectators make a spectacle of themselves: Or, social crisis and protonazi behavior come to the "Planet of the Apes"

"Hands up!" These are the words with which security guards greet people as they enter the Globe to view a contest between any two of the city's three major sporting clubs – AIK, Djurgården and Hammarby. A check of the body with a metal detector follows. No instruments which might bring injury to person or property are permitted. A precaution in keeping with previous "disturbances" at the Globe and Stockholm's various soccer stadia.

At one of the first events held at the Globe, an indoor soccer tournament, young fans supporting the three Stockholm teams clashed with one another on several occasions and in other instances destroyed a number of the arena's expensive seats.[197] At the same tournament in 1990, Hammarby supporters – a group highly displeased with the territorial incursions of the Globe[198] – went on a rampage, tearing up rows of seats, smashing six reinforced glass doors, casting potent firecrackers on the playing field and eventually precipitating a management decision to reject the tournament in future years.[199] Right up to the present, every ice-hockey match between AIK and Djurgården has required the presence of an extra contingent of police in and around the Globe in order to prevent their youthful fans – "the Black Army" and "the Blue Saints" – from scuffling or otherwise inflicting injury upon one another.[200] Tensions run high even on those occasions where violence is avoided, as the two groups – both predominantly short-haired and dressed in their required heavy boots and team colors – confront one another with threatening and taunting chants, with chants laden with symbolic creativity, with chants that merge the individual into the collective, that confirm identity through the confirmation of rivalry, that reaffirm the solidarity of otherwise disconnected and disoriented young men.[201] (The centrality of club affiliation to identity is captured by the following outburst of a Blue Saint: "Chicks and all that mean nothing, Djurgår'n is the only true love."[202]) These un-ruly circumstances and dis-orderly incidents have led one wit to nickname the Globe "the Planet of the Apes," a term which not only plays upon the "other wordly" spherical character of the Globe, but also upon "Monkey (or Ape) Mountain," the slang term for the standing area at the *Hovet* ice stadium where hockey matches were previously held.[203]

The raucous and sometimes violent behavior of the Black Army and the Blue Saints are neither spatially confined to the Globe nor phenomena unto themselves. The arena face-offs of these groups are usually preceded by several hours of public drinking, by getting "charged up" via copious beer consumption, by object- and fist-throwing skirmishes in pubs, by blood-drawing exchanges out on the streets. More of the same occurs after the matches. These "uncivilized," "unSwedish" manifestations of open conflict and aggression may be seen as symptomatic of something socially rotten in the state of Sweden (and other Western European countries). They may be regarded as the behavior of young men who are not infrequently economically and otherwise marginalized, who regard themselves as misunderstood by the authority figures in their lives and society at large, who sometimes see themselves as unfairly treated, unloved, unrespected and unsuccessful, who may be wracked with anxieties and a deep sense of alienation, who give meaning to their sense of social unacceptedness or outsidedness by acting socially unacceptably or outside the law. In its extreme form, perhaps present in a few per cent, the profile of these young men is identical with that of those who are members or sympathizers of White Aryan Resistance and other violence-prone racist groups operating in Sweden.[204] It is no mere coincidence that such neo-Nazi and "skinhead" groups are said to recruit among the Blue Saints and the Black Army.[205]

The glob(e)alization of Stockholm, Sweden, writ small

Another manifestation of glob(e)alization appeared in downtown Stockholm just two months before Globen City KB was caught in the quicksand of economic crisis, in a morass to which it had been a contributing agent.

MONDO – an exercise in creative destruction. The guts of which had been a branch bank totally ripped out – except for a downstairs vault – and restructured. Converted into an up-scale café, an assemblage of things and images, of commodified object/signs, that at one and the same time says Paris and anywhere in the world you wish to be.

MONDO – a miniaturized spectacular space of consumption. On the outside, tightly packed small round tables, a rolled up awning, and MONDO in can't-be-missed, glaring red letters. Inside the world as one big theme park. The alluring possibility of exotic experiences without ever leaving (Stock)-ho(l)me. An attempt to manipulate desire, to immerse the visitor in a set of expensive stage sets that speak directly to private fantasies, that signify the distant and thereby stimulate consumption.[206] Upon entering the customer is confronted by a wax-doll likeness of Humphrey Bogart and a huge travesty of a Henri Rousseau painting. Downstairs a bar is encountered with lavish wood finishings, marble flooring and a Louis Armstrong wax-doll looking down over it all. His gaze captures a bank of clocks set to commodified tropical escape spaces, to the local times of the Maldives, Phuket, the Great Barrier Reef, the Fijis, Hawaii, Puerto Vallarta and Miami. The adjacent wall is devoted to a mural of the globe encircled by commodification, by an equatorial halo reading *"La gente vino de la Oscurìda,"* by a ribbon spinning off into infinite space from the halo bearing a pastiche of clichéd images representing other commodified places – the Eiffel Tower, an Easter Island statue, the Coliseum, an Egyptian pyramid, an elaborately clad Indian elephant, coconut trees, an African woman bearing fruits and gourds on her head and so on. One upstairs "African" room contains lampshades of imitation leopard skin as well as tiger(!) masks and other "craft art" hung on the walls. Another such room is a cross between a private library and a well-appointed ship-cabin, is in either case a nook where the imagination is to venture where in the world it will. And, in order to transport the customer in time, in order to market "tradition," a throne atop the shelving behind the main-floor serving counter is occupied by a life-like wax figure of Gustav Vasa, heroic sixteenth-century king, founder of the modern Swedish state, precipitator of Sweden's Reformation.

MONDO is not, on the whole, simply a Baudrillardian "no-place space," simply a space of free-floating signs and decontextualized meanings.[207] It is instead an everywhere-at-once space, a cosmopolitan meeting place, a move-from-one-foreign-pavilion-to-another world's-fair-like phantasmagoric space. It is a space – like the Globe – where commodification and consumption are global projects, a space where a hypermodern world of unlimited possibilities (for consumption) is projected so as to seduce the visitor into seeking on-the-spot gratification. It is an evasion space, an enclave of experience removed from the tasks and worries of everyday life, a separate world where Sweden's globally linked unemployment and financial crises may be momentarily avoided.

MONDO is not, on the whole, simply a Baudrillardian world of simulacra, a world which is "hyperreal" because the "unreal" or "inauthentic" is made realer than real. It is instead a calculated world of (dis)simulation, where the real – the really produced by real people enmeshed in real social relations – is made recognizably more unreal than the unreal. Here, after all, the imitation Rousseau painting presents a landscape which is nothing more than an unrealistic version of *Le Douanier's* own haunting unreality. Here, after all, the unreal, surreal, juxtapositions of the barroom mural are made more unreal by one lower edge of it drifting off into a *coup d'œil* version of a bucket filled with dripping oil-paint brushes. Here, after all – thanks to special lighting and extra bread glazing employed in the interest of promoting sales, of perpetuating exchange and accumulation – the sandwiches and salads served appear redder than red, greener than green, shinier than shiny, and thereby more unreal than the plastic-unreal items often displayed in restaurant and foodstore windows.

On the whole, what is encountered is not simply a Baudrillardian world where there is "nothing behind the flow of codes, signs and simulacra,"[208] not simply a "postmodern" world devoid of political economy, materiality, difference, concrete historical reality and human agents capable of creating and reworking meaning. It is instead a **MONDO** space, where the fragmented and ephemeral conditions of flexible accumulation and hypermobile capital – of the contemporary global political economy – are mimicked and materialized.[209] It is a **MONDO** space where customers attempt to retreat from the historically unfolding and concrete ravages of glob(e)alization, from the experiences of a more than imag(e)inary political and economic crisis. It is a **MONDO** space where, in the course of chatter, women and men may creatively rework their mundane everyday worlds – elements of Stockholm's ongoing Europeanization/globalization, the blue sugar bags held in their fingers (bags decorated with the European Community's twelve yellow stars), the meaning of the repeatedly sighted Globe – and in so doing confirm local difference.

MONDO, a space where far from everybody is fooled by the illusory environment, where actively thinking women and men might be bemused by the figure of Gustav Vasa who, with his half-raised position and one leg moving forward in the air, appears to be saying: "Let me out of here! I've had enough of this unreally unreal place." A space where a reflective guest might observe "Everything is awfully nice here at Mondo, but, like, I don't feel at home in these surroundings."[210] A space where she and other like-minded persons might make wider crisis-laden associations leading them once more to ask:

What in the world is going on here?
Where in the world am I, are we?
Who in the world am I, are we?

How long before MONDO goes the way of Globen City KB?

How long before it proves just another mis-placed investment, just another short-lived by-product of Stockholm's glob(e)alization?

How long before it proves a financial disaster, writ small?

How many years or decades before the Globe itself goes belly-up economically, before its materialized metaphors and metonyms become anachronisms?

How many years before it proves no longer sustainable as a spectacular space of consumption, before it is bypassed in national and global circuits of accumulation?

How many years before it is ripe for creative destruction, before it is ready for the trash heap of history, before it is dis-placed by some new structure of capital?

How long before the revol(v)utionary nature of globally integrated capitalisms results in agents bringing forth yet another set of radical self-reinventions, results in the appearance of as yet unimagined products and forms of commodification?

How long before the need to market radically new forms of consumption, results in the construction of new spectacular spaces of consumption?

How long before the Stockholm Exhibitions of 1897 and 1930, as well as the Globe, are reechoed in greatly distorted, barely recognizable tones?

How long before Stockholmers, and Swedes in general, are confronted by yet another pivotal spectacular space of consumption, by a space that emerges out of and contributes to yet another set of (would-be) hegemonic discourses? How long before Stockholmers, and Swedes in general, are asked to accept a new set of meanings,
    to accept yet another remythologization of their past,
    to accept yet another simplified rewriting of their histories,
    to accept yet another reprogramming of their collective memory,
    to reidentify themselves along new lines of power?

How long before Stockholmers, and Swedes in general, having become dis-oriented, made to feel out of place, and having thereby grown discontent, once again tap into their deeply sedimented folk-humor traditions
    to culturally rework both their experiences of disjuncture and rupture,
    to culturally rework the shockingly new,
    to culturally rework both the new spectacular space of consumption and
        its associated would-be hegemonic discourses,

with or without political effect,
while simultaneously being driven to ask themselves,
yet again:
What in the world is going on here?
Where in the world am I, are we?
Who in the world am I, are we?

FINALLY: *POSTHISTOIRE(S)*, OR FUKUYAMA'S FINAL FOLLY?[211]

CARL HAMILTON and DAG ROLANDER (1992), Liberal-Party economists referring to the circumstances which forced Sweden to float the *krona*, to relinquish its adherence to a fixed-link exchange rate and the "free flow of capital"

[Conservative Prime Minister] Carl Bildt has now encountered the market and he doesn't approve of what he saw.

PERRY ANDERSON (1992), 335 (emphasis added)

Hegel's theory of the State had envisaged a synthesis of freedom and identity – self-determination as both representation and as expression. What comparable moral substance does the contemporary political order in the West have to offer? The most frequent liberal response today is to dismiss the question as misplaced: in a democratic society, the public arena *is necessarily no more than the instrumental space in which substantive private goals of diverse kinds may be pursued.*

From childhood reminiscences contained in INGEMAR UNGE (1992)

[In our imaginations] one of the ["Lizard"] ponds {now buried by Globen City} was the eye of history, a tunnel to other times, a mirror of that which had already happened, a gateway to the DDR, De Dödas Rum [the Chamber of the Dead], a telescope through which one could see down into Hades.

HANS KÄLLENIUS, as quoted in *Dagens Nyheter*, May 29, 1993

That project [an international exhibition to commemorate the centennial of the Stockholm Exhibition of 1897] fits extremely well with my post as managing director of Stockholm Propaganda Inc.... [It's] a chance to place Stockholm on the world map and give Sweden a flying start in the new Europe.

FIGURE ACKNOWLEDGMENTS

Photographs of the Globe by kind courtesy of Martin Nauclér and *Arkitekturmuseet* (The Swedish Museum of Architecture), as well as Upside AB (aerial shot including the Stockholm Globe Hotel). Photograph of Gunnar Asplund's planetarium also by courtesy of *Arkitekturmuseet*. People's Hall photograph from Roger R. Taylor (1974), *The World in Stone*, by courtesy of the University of California Press. The graffiti art was found, copied and generously provided by Barbro Klein. The cartoons, by Jonas Darnell and Patrik Norrman, are reproduced from *Dagens Nyheter*, August 18, 1991.

NOTES

1. Sweden's role as a "Great Power" in Europe is usually regarded as having extended from 1611 to 1718.
2. The Debord quote is from thesis no. 34 of *La Societé du Spectacle*. The rendition provided by Marcus is less clumsy than that provided in the standard English translation (Debord, 1983).
3. Roger Miller (1992).
4. Of course, semi-globular, or domed buildings have a much longer history. In the western archtitectural tradition cupola-topped buildings, such as the Pantheon in Rome, date back at least to classical antiquity.
5. Reutersvärd (1989b).
6. The diameter for each of Boulées temples exceeded that of the Globe. For illustrations of these and other globular schemes by Boulée's contemporaries see Björkman (1989), 99.
7. Torsten Ekbom in response to Oscar Reutersvärd (1989a). Ekbom and Reutersvärd engaged in a debate, occasioned by the Globe's opening, as to whether or not the purist geometrical projects of Boulée and his confreres may be regarded as a form of pioneering modernism, as involving the same radical erasure of all traces of ornamentation as the Constructivism associated with the Russian Revolution.
8. Dunbar (1974), 57, 59.
9. Benedict (1983), 56.
10. See Dunbar (1974, 1978) on the ill-fated plans of Elisée Reclus, the French geographer and anarchist, for a relief globe that would dwarf Villard-Cotard's and serve as a centerpiece for the 1900 Paris *Exposition Universelle*.
11. Sorkin (1992a), 225.
12. Bennett (1988), 96–8.
13. Björkman (1989), 100; Dunbar (1974), 63. The spherical attributes of the Prague gas-holder, which rested on eight concrete feet, also enabled a larger area to be served.

14 Buck-Morss (1989), 315–16, 471, citing Taylor (1974).
15 Sorkin (1992a), 225.
16 Björkman (1989), 101.
17 Björkman (1989), 102.
18 Cf. Watts (1991) and Pred and Watts (1992). Also note Aggar (1989), Harvey (1989b), Henderson and Castells (1987) and Virilio (1988).
19 Storper and Walker (1989).
20 Thrift (1992), 4. For details see Thrift in Thrift and Corbridge (1994).
21 Such contracting has become pervasive in the realm of agriculture as well as manufacturing. See Watts in Pred and Watts (1992) 65–83, and Watts and Little (1994).
22 "The difference between living standards in Europe and in India and China increased from a ratio of 40:1 to 70:1 between 1965 and 1990" (Anderson, 1992, 353). For details see Arrighi (1991).
23 Featherstone (1991), 110.
24 Stuart Hall (1991), 27. Of course, this is not to suggest either that these images were everywhere given the same meanings, or that they were always – or even for the most part – received passively, without cultural reworking.
25 Cf. King (1992), 84–5, and the literature there cited.
26 Featherstone (1991), 109.
27 Cf. Harvey (1988), (1989a), (1989b), 92; Britton (1991), 467–70; Zukin (1991); Knox (1992), 10–12.
28 Rich and extensive childhood reminiscences about the area are contained in Unge (1992).
29 This account, and that of the remainder of this section, is based primarily on Sahlin-Andersson (1989). Additional details have been obtained from AB Stockholm Globe Arena (1989), Torbjörn Andersson *et al.* (1991) and numerous editions of Stockholm's two leading dailies, *Dagens Nyheter* and *Svenska Dagbladet*, which are not cited except where quoted.
30 Cf. Roger Miller (1992).
31 Sahlin-Andersson (1989) 131, quoting the announcement for the project competition mentioned below.
32 Twenty-six million crowns were well under four million U.S. dollars at then prevailing exchange rates.
33 The almost doubling of the office-space quota was not the only concession granted by the municipality's negotiators. As originally conceived, the group winning the competition would be given only a leasehold, rather than a freehold, on the land they were to develop. Moreover, by 1990 the consortium was permitted to readjust the amount of usable office, shop and hotel space – it would finally construct up to 163,000 square meters.
34 This was another concession meant to compensate for the loss of office rental income resulting from the requirement that the Globe was to be completed before any other complex building.
35 Neither the type of accommodation reached between agents of capital and the

local state with respect to development of the *Johanneshov* area, nor the extraordinary procedures associated with its implementation, are fully peculiar to Stockholm. While their specifics may be unique, they are symptomatic of a more widespread capitulation of local government to "market forces" in conjunction with the development of public works and spectacular spaces of consumption. Urban "material landscape[s] created by the joint efforts of speculative developers, elected officials, financial institutions, and architectural designers" have become commonplace in the U.S.A. (Zukin, 1991, 54). There, public–private partnerships have wed local "crisis" management and speculation to produce "a rapid proliferation of spectacular set-piece projects ... , Baltimore's Harbor Place, Riverwalk in New Orleans, Riverfront in Savannah, Quincy Market in Boston, Pioneer Square in Seattle, South Street Seaport in New York" and so on (Knox, 1992, 11; cf. Harvey, 1989a, 1989b). Likewise, in Sydney (Darling Harbour), in Edmonton (the West Edmonton Mall), in Toronto (Harbourside), in London (Covent Garden, St Katharine Docks), in Paris (the former *Les Halles* district) and elsewhere in Europe, there has been joint public–private participation in the development of spectacular "festival market-places" and consumption complexes (cf. Britton, 1991, 472; Crawford, 1992, 28).

36 There was very little in the way of open mass-media criticism of the Globe project throughout the period it was taking shape and being negotiated. In an atmosphere in which a number of powerful political and economic interests had sung of the entire enterprise in a unified chorus of glowing positive terms – in terms of its grandeur, its spectacular qualities, its boost to culture, its profitability and effectiveness – there were few public figures who wished to appear as if they were getting in the way. The limited amount of criticism making its way into the press mostly centered on such questions as the shape and scale of the Globe and its impact on the skyline. Such commentary functioned, in essence, as an aesthetic smokescreen for other potentially sensitive issues involving the remainder of the project, the undermining of democracy and Social Democratic complicity in the privatization of Stockholm real estate at the expense of housing development (cf. Sahlin-Andersson, 1989; Torbjörn Andersson *et al.*, 1991; and Roger Miller, 1992). Mass-media coverage assumed more diverse critical forms from 1989 onward – after the Globe's opening – when the actual cost of the project to taxpayers and the municipality became a matter of controversy, especially after Globen City KB demanded 250 million crowns of authorities for extra costs incurred as a result of design and standard modifications imposed during the course of construction.

37 *Dagens Nyheter*, October 28, 1989.

38 Mats Hulth, Social Democratic municipal council member, as quoted in *Dagens Nyheter*, February 16, 1989.

39 Roger Miller (1992). It is to be recognized that the "City of Stockholm held (and still holds) a near monopoly on all developable land, as well as veto power over the kinds and design of projects for which land can be used" (Miller, 1992).

40 Mats Hulth as once again quoted in *Dagens Nyheter*, February 16, 1989.
41 In 1989 the Globe was awarded an architectural prize by the European Steel Convention, a group which previously had granted the same distinction to *La Pyramide du Louvre* in Paris (*Dagens Nyheter*, November 8, 1989).
42 Sahlin-Andersson (1989), 138.
43 AB Stockholm Globe Arena (1989), 2.
44 In 1985 and 1986, Stockholm respectively ranked fifteenth and eighteenth in the world in terms of international conventions held (King, 1992, 91, citing Knight and Gappert, 1989, 323).
45 Ironically, during the period of the Globe's development, Stockholm-based investors, acting as agents in the international circulation of capital, were becoming ever more feverishly engaged in real-estate speculation, office-space development and the acquisition of highly visible "trophy buildings" in both London and Brussels. Cf. King (1992), 98.
46 A popular nickname for the skyscraper housing the entire administrative apparatus of the tax authorities for the province, or county, of Stockholm.
47 Cf. Harvey (1989b, 83) on both the schizophrenic qualities of postmodern architecture and Jencks' version of the "double coding" that "must" be embodied in such architecture.
48 Cf. "A globe for sport and culture," in AB Stockholm Globe Arena (1989), 5.
49 Doordan (1988), 134. Cf. Gargas (1981); Etlin (1991); and Frampton (1992), 203–9. Terragni's *Casa del Fascio* is generally regarded as "the canonical work of the Italian Rationalist movement" (Frampton, 1992, 205). I wish to thank Carol Krinsky, the NYU art historian, for initially informing me of the connection between the Stockholm Globe Hotel and Terragni's Como structure.
50 Giuseppe Terragni, "La costruzione della Casa del Fascio di Como," *Quadrante*, no. 35/36, as quoted in Doordan (1988), 137.
51 Hughes (1991), 99. Regarding the Corso, see the previous chapter, pp. 112–16. Here there may be an additional oblique reference to the Stockholm Exhibition of 1897, as Ferdinand Boberg, the architect behind the Hall of Industry and other key exposition buildings, is known to have taken pride in the audience given him many years later by Mussolini.
52 Cf. Roger Miller (1992), who refers to the specialty meat stores and other small-scale establishments once clustered around the entrance to the stockyards and slaughterhouse district.
53 Östlundh (1989), 96.
54 Samec (1989). All preceding terms in quotation marks are from this source, which is based on statements made by representatives of *Berg Arkitektkontor AB*.
55 Werckmeister (1991), 12–13.
56 Harvey (1989b), 91.
57 Debord (1971).
58 This is not a pejorative hurled from the left, but a long-standing common usage as well as a term of self-designation used by the parties in question.
59 A slogan deployed by the Social Democrats during the 1991 parliamentary elections.

60   Crayfish consumption in Sweden, which occurs in August, is surrounded by a number of traditional rituals, by a set of traditions of more than a little symbolic significance to national identity. Sweden's crayfish have been greatly decimated by disease and the country has become increasingly dependent on frozen imports from Louisiana, California and elsewhere.
61   Cf. Watts (1991), Pred and Watts (1992) and the sources cited therein. Also note Hoggett (1992).
62   Summary of a 1988 telephone interview with the information director of AB Stockholm Globe Arena as reported in Torbjörn Andersson *et al.* (1991), 2.
63   Torbjörn Andersson *et al.* (1991), 3, citing *Dagens Nyheter*, June 3, 1986.
64   On the "capitalist penetration of popular-music production," the "inexorable" commodification of music which "expresses opposition to the social order," and the globalization of the music industry see Bloomfield (1991).
65   The rink boards surrounding the ice-hockey players are also plastered with the advertisements of firms that are in one way or another linked into the global capitalist economy. Some ice-hockey players also carry advertisements for various bookmaking and sports-results betting pools on their uniforms, thereby making themselves commodified bodies advertising for enterprises that in effect commodify other bodies. And, of course, photographs of prominent players regularly appearing at the Globe are used for the promotion of goods via mass-media advertisements.
66   Cf. Buck-Morss (1986), especially 121; and Wilson (1992), 105–7.
67   *Dagens Nyheter*, January 21, 27, 1992. Subsequently revised figures indicated that the Globe just about broke even in 1991. The total attendance for 1991 (1,235,490) was greater than that for 1990, but below the opening-year figure (1,384,249) (AB Stockholm Globe Arena, 1992).
68   All information on suppliers derived from Karlström (1989), 14–15, 106–7 and AB Stockholm Globe Arena (1989), 5.
69   Harvey (1989b), 142, 147, 171; emphasis added.
70   Architect Svante Berg as quoted in *Dagens Nyheter*, February 16, 1989.
71   The track-and-field configuration also lends itself to equestriatian show-jumping and motorcross competitions.
72   Östlundh (1989), 94.
73   The Swedish government formally applied for European Community membership on July 1, 1991. The Social Democrats, who were then still in power, first revealed such an intention during October 1990, in connection with a package of "economic-crisis" proposals.
74   As of mid-1992, at least, Swedish membership in the European Community was supported by the leadership of all but one of the seven parties sitting in Parliament, even if the pro-membership faction of the Social Democrats was increasingly unclear and somewhat reluctant to use explicit "free market" language. Unequivocal opposition came from the Left Party (which had previously included the designation Communist in its label). Strong public dissent also was voiced by the Environment (Green) Party, which had dropped

out of Parliament after failing to obtain 4 per cent of the vote in the 1991 election.
75 Lodin (1992). Cf. the epigram by Göran Tunhammar, p. 178, above.
76 Between 1985 and 1991 the total number of manufacturing workers employed in Sweden by domestic *and* foreign firms fell from 957,000 to 905,000. During the same period the number of new industrial laborers employed at overseas locations by Swedish corporations grew from 329,000 to 615,000 (Liljefors and Bergquist, 1992). In 1990 alone, Swedish industrial groups invested 80 billion crowns abroad, while foreign corporations made investments within the country that only amounted to a tenth of that sum (Swedish Trade Council, 1992, 27).
77 Lodin (1992).
78 Regarding the "Great Consumption Party," see p. 212.
79 The intensity and breadth of this questioning increased markedly in June 1992, after the first Danish public referendum on the Maastricht Treaty, which resulted in a temporary rejection of that document. These and subsequent questions in this section are generalized reformulations based largely on unstructured informal conversations participated in, and overheard in public places, during the sixteen months I spent in Sweden between December 1990 and August 1992. These conversations, which came up in the conduct of everyday life, involved people of both genders who varied widely in terms of their age, occupation, class background and political allegiance. These questions are also partly derived from my intense daily exposure to radio and television reportage and interviews, as well as from the press. These enmeshed observations were facilitated by earlier research stays in Sweden totalling over eleven years.
80 Andersson and Edin (1992).
81 This question should not obscure the fact that most Swedes are rather pleased, if not gleeful, at the prospect of cheaper and more widely available alcoholic beverages.
82 Wording of the Maastricht Treaty, as translated from the Swedish version of this phrase repeatedly quoted in the daily press.
83 Cf. Carlberg (1992).
84 Some of the more economically sophisticated also questioned whether or not a monetary union would force the wages and prices of different countries to converge, thereby masking differences in productivity and spelling doom for companies and peripheral regions characterized by low productivity (e.g., Sandberg, 1992).
85 Cf. Andersson and Edin (1992). The survival of democracy within Sweden has also come into question in some circles because of the fate of funds put aside by parliament for the dissemination of pro- and con- information regarding the The European Community. The 50 million crowns allocated were supposed to be equitably distributed, with the idea that the population would be prepared to vote for or against the matter in a national referendum once Sweden was actually offered membership. By August 1992, the government had appro-

priated 35 million of the total for itself, claiming that much of that sum was to be used for providing "basic information" in a series of purely "objective" and cheaply priced "green books," and for translating various documents rather than the spread of arguments favoring membership (*Dagens Nyheter*, August 4, 1992). Meanwhile, the remainder was to be competitively awarded to organizations through a commission headed by a former Prime Minister, Torbjörn Fälldin, whose Center Party belongs to the ruling bourgeois-block coalition headed by the Conservative Party. (However distortive of democratic principles this appeared to some, they would have had to admit that support for membership was far from unanimous within Fälldin's party.)

86 Cf. Sörlin (1991); Schlesinger (1991); and Daun (1992).

87 Regarding the construction and elements of regional, as opposed to national, identity in the contemporary world, see Stuart Hall (1991), 35–6; Paasi (1986, 1991); Pred (1983); and Bourdieu (1991), 220–8. Some of Giddens' observations (1991) on "ontological security" and self-identity may be read into the formation of regional identity.

88 These questions were, of course, not peculiar to Sweden, but pandemic to an entire continent given the ongoing transformations of Eastern Europe and the former Soviet Union, the debates about European Community membership simultaneously occurring in Austria, Finland, Norway and Switzerland, as well as the moves toward greater economic and political integration within the European Community itself. Similar questions commonly arose elsewhere in the world, especially in conjunction with the local introduction of new forms of capital (Pred and Watts, 1992).

89 From "*Sveriges mest spännande hotel*" (Sweden's Most Exciting Hotel), a brochure distributed by the Stockholm Globe Hotel. In their newspaper advertisements the hotel's operators (Best Western) repeatedly allude to the luxury quality of the Arena Restaurant, to food which matches "world-class" entertainment.

90 Alcohol consumption also occurs behind the glass façades of the luxury boxes. Class segregation at the Globe is further underscored by a "V.I.P." entrance, officially designated as the "Golden Gate."

91 As the writings of Benjamin make evident, the (con)fusing of public and private urban space via commodification dates back to the nineteenth century (Buck-Morss, 1989; Wilson, 1992). Cf. Zukin (1991), especially 51–2, and the papers contained in Sorkin (1992b).

92 Significantly, it is only since the opening of the Globe, since the widespread preaching of the new free-market gospel, that Swedish households have had access to advertisement-carrying television channels.

93 Informant (male, civil servant, 25-years old), EMSU.

94 Cf. Zukin (1991), whose use of "liminality" with respect to urban landscapes is derived both from Turner's original usage (1969, 1982, 20–60) and its later extension to market phenomena by Agnew (1986).

95 The deployment of these various cameras was approved in each specific instance by either provincial or national authorities (*Dagens Nyheter*, January 12, 1990).

96 From "Ideas for Our Future," a program outline issued by the Moderate (Conservative) Party in 1990.

97 Only one year after its election, the neoliberal "bourgeois coalition" had lost much of its voter support. Public opinion polls showed that the four coalition parties were together favored by only 37.5 per cent of those willing to express a view (9.5 per cent of those surveyed could not or would not indicate a preference). Support for the Social Democrats had shot up to 47.5 per cent, while another 5.0 per cent displeased with the government's policies supported either the Left or Green Parties. At the same time, 9.0 per cent chose to back the highly populist New Democracy Party (*Dagens Nyheter*, October 17, 1992).

98 Cf. note 80, above. According to a fall 1992 estimate, 35,000 jobs would alone disappear from hospitals and clinics over the next four years (*Dagens Nyheter*, September 26, 1992).

99 Widespread concern led to the organization of a national "Children's Lobby" which was to fight against any reduction in expenditures for child-care centers and schools, and was to defend the appropriations made for libraries, museums and other cultural activities oriented toward children.

100 In no small measure owing to the creation of public-sector jobs in recent decades, 84 per cent of all Swedish women aged 20–65 were gainfully employed as of 1991 (Swedish Trade Council, 1992, 260).

101 Lindqvist (1992).

102 *Dagens Nyheter*, August 6, 1992.

103 *Dagens Nyheter*, June 30, 1992.

104 This is not merely a popular view held in a variety of circles, but also the view of Kjell-Olof Feldt (1991), who was the Social Democratic Minister of Finance during the period in question.

105 From the end of 1986 through early 1990 the average Swedish household had a negative savings quotient, or expenditures that exceeded disposable income. At the low point in this trough, the average household was annually spending five per cent more than it had available (data gathered by the Central Bureau of Statistics (*Statistiska Centralbyrån*) and the Ministry of Finance (*Finansdepartementet*) and reproduced in *Dagens Nyheter*, October 24, 1992).

106 Cf. Feldt (1991), 281–3. This is not to suggest, of course, that the eventual decrease in Swedish manufacturing employment was solely a consequence of the "Great Consumption Party." The eventual crisis of manufacturing (and other sectors) also can be traced in large measure to the massive detouring of funds into short-term speculative ventures and overseas industrial investments, especially in European Community countries and the "Third World."

107 See Löfgren (1992, source of quote) regarding the aesthetics of contemporary Swedish consumption, especially with respect to household furnishings, which were an important element of the "Great Consumption Party." (According to Löfgren's sources, Sweden has Europe's highest per capita consumption of household furnishings and decorations.)

108 Göransson (1989).

109 Cf. Foster (1985) on the spectacle as commodity.
110 The advertisement ran full page in a number of large-circulation newspapers on January 29, 1989. That the commodified levelling of "high" and "low" occurring in the Globe is symptomatic of a more widespread refusal to draw commodity distinctions is suggested by the efforts of a leading advertising agency to advertise itself: "Our job is to get things to happen. A government can, for example, ask us to help them to remain in power.... A large international corporation requests us to develop concepts for the test launching of a completely new product on the European market.... An erotic magazine asks us to help them put some zip in their circulation" (*Svenska Dagbladet*, September 1, 1991).
111 *Dagens Nyheter*, April 12, 1992.
112 *Dagens Nyheter*, June 22, 1992.
113 For specification of these revelations, see p. 218.
114 Some on the left who opposed membership in the European Community have argued the opposite, that Sweden has never been a part of Europe. However, see Sörlin (1991).
115 Cf. Featherstone (1991), 78–82, on the "controlled de-control of the emotions" occurring in modern spectacular spaces of consumption, and its relation to the "carnivalesque" as developed by Bakhtin (1984) and Stallybrass and White (1986).
116 *Dagens Nyheter*, February 16, 1991.
117 *Folket* (Nyköping and Eskilstuna), September 18, 1991.
118 *Dagens Nyheter*, March 12, 1990.
119 Wording of the Maastricht Treaty, as translated from the Swedish version of this fragment repeatedly quoted in the daily press.
120 The "notice" was to serve as a foundation for the negotiation instructions to be developed by the European Community's Council of Ministers in the fall of 1992.
121 The Commission's "notice" stated that: "An appraisal of the relevant articles in the Maastricht Treaty and [t]herewith associated declarations leads to the conclusion that Sweden should be able to fulfil all obligations following from a common foreign and security policy." It was apprehensively added, however, that Sweden might act as a brake "on the possible development of a common foreign and security policy" (translation of the Swedish text as quoted by Tarschys, 1992).
122 Cf. Daun (1989), 102–23, and the literature cited therein.
123 Bildt (1992).
124 Åström (1992).
125 *Dagens Nyheter*, July 4, 1992.
126 Lodin (1992).
127 For example Tarschys (1992). Now that there was no longer an anti-West, pro-Soviet "non-alliance" movement led by the likes of Tito, Nehru, Castro and Sukarno, the term could be used comfortably by members of the self-styled bourgeois parties.

128 Cf. note 80, above.
129 Membership supporters emphatically replied in the negative, arguing this wouldn't be possible because all military actions will require an unanimous vote of the member states.
130 Seeds of doubt were also cast by the retelling of the tale of Olle Hedlund, an author who was blacklisted for more than a decade after his satirical anti-Nazi radio-play, "Blondes" ("*Blondiner*"), was cancelled by the Swedish Broadcasting Corporation (*Sveriges Radio*) in 1939.
131 *Dagens Nyheter*, May 27, July 27, 1992.
132 Friggebo and Westerberg (1992); *Dagens Nyheter*, August 8 and 22, 1992. Correspondence from Roger Andersson, Uppsala University.
133 Prior to a policy turnaround in late 1990, the Social Democrats had long prided themselves in their extremely generous position, which allowed people with "refugee-like" circumstances to be permanently admitted to the country. The sudden retreat to the Geneva Convention definition occurred in the context of a rapid increase in the number of residence permits granted to refugees between 1983 (5,000) and 1989 (25,000), a mounting difficulty in finding housing for those people (most of whom were Muslims arriving from the various "trouble spots" of the Middle East), and a new upsurge in the number of asylum applicants assigned to widely scattered camps where they were to wait as long as two years for a decision on whether or not they could remain in the country. The Social Democrats justified their reversal, at least in part, with the argument that Sweden would sooner or later become a part of the European Community and therefore would have to adopt a refugee policy in accord with that of most members.
134 In addition, the 3,800 Macedonians who arrived in a single week were sent back immediately, having been classified as economic, rather than political, refugees (*Dagens Nyheter*, July 22, 1992). The Social Democrats had set a precedent for turning away refugees in large numbers during 1989, when the door was shut to thousands of fleeing Bulgarian Turks. As early as 1990, Swedish authorities were granting residence permission to only 43 per cent of all refugee applicants; a figure that was low by previous Swedish standards but high by prevailing European standards (*Dagens Nyheter*, July 8, 1992).
135 *Dagens Nyheter*, August 8, 1992.
136 Friggebo and Westerberg (1992).
137 *Dagens Nyheter*, July 10, 1992.
138 These statements were usually coupled with reminders that Sweden's per capita receipt of refugees surpassed the rate of all other countries. The fact that the United States had all but locked out refugees from the former Yugoslavia was rarely mentioned by government politicians, in part, presumably, because of their ideological reluctance to criticize their conservative model across the Atlantic.
139 On different occasions the cabinet ministers most involved, as well as the Prime Minister himself, referred to this alternative as "rational," "humane" and

"cheaper" – involving about one-fifth the cost per refugee-day (Friggebo and Westerberg, 1992; *Dagens Nyheter*, July 31, 1992). At a UN emergency conference held in Geneva, Minister of Immigration Affairs Friggebo also proposed that Sweden and other countries might deal with the Bosnians by offering them temporary residence permits. This, she believed, would encourage them to remain in "neighboring territories" (*Göteborgs Posten*, July 29, 1992).

140 *Dagens Nyheter*, July 11, 22, August 22, 1992; numerous Swedish radio and television broadcasts during July and August 1992.

141 Georg Andersson (1992).

142 Study cited in *Dagens Nyheter*, June 26, July 5, 1992.

143 For an account and analysis of such jokes, rumors and anecdotes as they existed in Sweden during the 1970s see af Klintberg (1988) and Velure (1988). Such bits of folk "humor" began to be all the more common during the 1960s when labor shortages resulting from the expansion of Swedish manufacturing and concomitant increases in the globalization of Stockholm brought a large influx of Finns, Yugoslavs, Greeks and Turks.

144 Harry Schein – long-resident immigrant, entrepreneur and confidant to prominent Social Democrats – commenting on the spread of racism in Sweden (Swedish Trade Council, 1992, 29).

145 *Södermalm* is a large island and a traditional working-class district that has been considerably transformed in recent years by widespread gentrification and the architectural post-modernization of an extensive railyard area (*Södra Station*). These transformations, like the Globe City project, were symptomatic of the often wildly speculative real-estate developments occurring in Stockholm during the 1980s.

146 Cf. Willis (1990) and Daniel Miller (1987) on the work of symbolic creativity associated with consumption objects.

147 In Swedish the definite form of a noun is arrived at by attaching its indefinite article on to its end.

148 Sahlin-Andersson (1989), 145–6; Roger Miller (1992). English terms are often punned upon or otherwise employed in Stockholm folk humor.

149 From an informant (female, office worker, 50 years old), EMSU.

150 From an informant (male, student, 25 years old), EMSU.

151 From an informant (male, ombudsman at the Swedish Union of Commercial Employees and goalkeeper for a Stockholm hockey team, 26 years old), EMSU. This informant claimed there was no slang term for the Globe among professional and semi-professional hockey players. *Snorkråkan* may be more literally translated either as a clump of snot that flies like a crow (*kråka*), or as a clump of snot shaped like an editor's tick mark, or editorial dot (also *kråka*).

152 This was an image projected by local business and political leaders, including the top executives of Volvo, which is headquartered in Göteborg. They contended, among other things, that Göteborg was closer to the continent than Stockholm as well as being half way between Copenhagen and Oslo. The very name Scandinavium is indicative of the center-of-the-Scandinavian-universe

investment-attracting image they were attempting to project.
153 From an informant (male, salesman, 27 years old), EMSU.
154 *Arbetarbladet* (Gävle), February 5, 1989. The literal translation for *pratbubblan* is "the talk(ing) bubble."
155 *Dagens Nyheter*, December 7, 1989.
156 *Svenska Dagbladet*, April 8, 1989; *Aftonbladet*, February 2, 1989.
157 *Aftonbladet*, February 2, 1989.
158 See pp. 189–90.
159 *Dagens Nyheter*, December 22, 1990. Those thus disturbed, as well as others, were not made happy by news that the consortium operating Globen City was planning a sharp rent increase for the municipally operated nursery to be opened for children of those employed at the office and consumption complex (*Dagens Nyheter*, February 21, 22, 1990).
160 *Dagens Nyheter*, May 6, 1989.
161 From an informant (male, order-placement clerk, 25 years old), EMSU.
162 *Dagens Nyheter*, February 5, 1989.
163 Torbjörn Andersson *et al.* (1991), 18; *Dagens Nyheter*, June 30, 1989. Those protesting over the pond also claimed it was the home of a rare fish variety (Sahlin-Andersson, 1989, 202). Those registering a complaint over the trees asserted that their root systems did not meet legal standards.
164 *Dagens Nyheter*, February 26, 1989. The tabloid press broadcast this occurrence as a "scandal" (*Expressen*, February 26, 1989). In order to quiet the media storm, the firms in question quickly announced they would refund the small purchase price (five crowns) to each person returning a ticket (*Dagens Nyheter*, February 28, 1989).
165 Letter from a politician quoted in *Dagens Nyheter*, June 13, 1989.
166 Cf. Bakhtin (1984, 458–63) on "praise-abuse" and nicknames. Also note Morgan, O'Neill and Harré (1979), Stallybrass and White (1986), 44–59 and Pred (1990a), 144–85.
167 From an informant (female, student, 22 years old), EMSU; and *Norrköpings Tidningar*, April 1, 1989.
168 From a personal informant (male, retail store employee in the Globen City shopping mall, 23 years old).
169 From several informants (female, businesswoman, 45 years old; male, salesman, 27 years old; female, student, 34 years old; female, office worker, 50 years old), EMSU; *Dagens Nyheter*, March 9, 1991 and January 18, 1992; and *Svenska Dagbladet*, September 25, 1992. According to two apparently apocryphal accounts, architects first got the idea for the Globe from a chance encounter with a golfball (Sahlin-Andersson, 1989, 60).
170 From two informants (male, doctor of Chilean birth, 37 years old; woman, student, 22 years old), EMSU; and *Dagens Nyheter*, January 24, 1992. The light emitted from the Globe's distinctive small windows in times of darkness has probably contributed to the spaceship and moonbase associations. Also note mention of "the Planet of the Apes," p. 236.

171 *Dagens Nyheter*, January 18, 1992.
172 From an informant (male, student, 23 years old), EMSU.
173 From several unspecified informants, EMSU.
174 From the same unspecified informants of the previous note.
175 This kind of symbolic reworking of discontent had a counterpart in the name attached to a skyscraper that was central to the controversial "postmodern" renewal of Stockholm's *Södra Station* area (see note 146, above). "[T]he tower was derisively christened *Haglund's Pinne*, after Sune Haglund, a Conservative Real Estate Director" (Roger Miller, 1992, 18). This was a doubly abusive nickname, as it suggested that neither the building itself, nor the "manhood" of a powerful figure was anything more than a slender stick or twig (*pinne*).
176 Olsson (1991b), 17–18. In further playing upon the phallic, Olsson suggests that contemporary planning in Sweden (and elsewhere) results in "a modern version of the castration complex in which some [the power holders] fear the loss of something they once had, while others [the power subjects] experience the lack of something they never possessed." *Josefsson's ollon* might be seen as at least a subconscious re-cognition of this circumstance.
177 Cartoon series by Jonas Darnell and Patrik Norrman, *Dagens Nyheter*, July 28, August 4, 11, 18, 25, 1991.
178 Klein (1992).
179 The Swedish Trade Council is a collaborative venture between major Swedish corporations and the State. It operates thirty-six offices around the world.
180 In 1991, the Swedish gross national product shrank by approximately 1.5 per cent (Swedish Trade Council, 1992, 15). It continued to decline in 1992 and was projected to diminish 1.0–1.5 per cent during 1993 (*Dagens Nyheter*, October 24, 1992).
181 By September 1992, unemployment had reached 5.2 per cent and there were prospects of the figure soon rising to 7.0 per cent or more (*Dagens Nyheter*, October 10, 24, 1992). (Although these would be regarded as "tolerable" levels in most industrialized countries, they were perceived as astronomical in a country accustomed to structural overemployment [a greater number of unfilled jobs than unemployed workers]). In fact, by the summer of 1993 the unemployment level had reached nearly 11 per cent and another 3 per cent of the work-force were only technically employed, being engaged in government subsidized job-training programs or public works projects.
182 In 1991 Swedish industrial output fell 5.5 per cent after dropping 2.0 per cent in the previous year (Swedish Trade Council, 1992, 5). Even factories that were not suffering from declining sales were often subject to employment reductions owing to the application of "lean-production" and other cost-cutting strategies.
183 In order to stem the movement of currency-speculation funds out of the country by Swedish corporations, banks and insurance companies, marginal interest rates on such transactions were set at 500 per cent by the Central Bank of Sweden (*Riksbanken*) on September 16, when it had become evident that German interest rates were not coming down and the British pound was being

allowed to float freely. Before the end of November pressures proved too great and the krona was allowed to float against the German mark and the ECU (European Currency Unit), and thereby against other major currencies.

184 At then prevailing exchange rates, those debts were the equivalent of well over $300 million.
185 *Dagens Industri*, February 23, 1989.
186 *Svenska Dagbladet*, September 25, 1992.
187 *Dagens Nyheter*, September 25, 1992.
188 The third and smallest phase of construction, which involved only 9,000 square meters of office space, was at the time well under way. With the announcement of the payment stoppage further construction was immediately halted. Cf. note 33, above, and the text thereto.
189 Principally owing to bad real-estate loans and the collapse of real-estate prices, the Stockholm-centered Swedish banking system had absorbed credit losses of about 30 billion crowns in 1991 and was still bleeding profusely in the fall of 1992 (Swedish Trade Council, 1992, 15). In fact, total credit losses for all of 1992 turned out to be at least twice those of 1991, and by the beginning of 1993 five of the country's six major banks were receiving some form of financial assistance from the State (*Dagens Nyheter*, January 9, 1993).
190 AB Stockholm Globe Arena (1992), 11; *Dagens Nyheter*, January 22, 1992.
191 AB Stockholm Globe Arena (1992), 10; *Dagens Nyheter*, January 22, 1992.
192 Instead, in May of 1993, an agreement was reached whereby the annual rent paid by Stockholm Globe Arena Inc. was to be successively decreased over a ten-year period from 15 million crowns to zero. In return, the municipality would no longer be required to make contributions in order to help cover losses. (Despite other financial pressures, the muncipality had been forced to contribute 11 million crowns for 1992 [*Dagens Nyheter*, May 29, 1992].)
193 Cf. note 90, above.
194 Commentary in *Dagens Nyheter*, January 24, 1992, under the headline "Global cheat-lunch."
195 AB Stockholm Globe Arena (1992), 5; *Expressen*, August 22, 1991.
196 *Dagens Nyheter*, August 25, 1991; *Expressen*, August 22, 1991; *Aftonbladet*, October 2, 1991.
197 *Kvällsposten*, February 23, 1989; *Smålandsposten*, February 23, 1989.
198 From an informant (male, order-placement clerk, 25 years old), EMSU. Cf. p. 225, above.
199 *Dagens Nyheter*, January 25, 26, 1990.
200 Hammarby suporters are unmentioned here, as the club is not a member of the top ice-hockey division and is consequently forced to play its matches elsewhere.
201 Cf. Willis (1990, 109–15) on young, white, male working-class football fans in England. That AIK and Djurgården supporters in some measure model themselves after their rough-and-tumble English counterparts is suggested by their self-chosen English-language collective labels.

202 *Svenska Dagbladet*, February 1, 1991.
203 *Aftonbladet*, December 8, 1991.
204 Cf. Daun (1992), 196–201.
205 Daun (1992), 200; *Svenska Dagbladet*, February 1, 1991.
206 Cf. Boyer (1992), especially 200, 204.
207 Cf. Featherstone (1991), 99.
208 Kellner (1989), 83.
209 Cf. Harvey's arguments (1989b) on how the conditions of flexible accumulation are mimicked in various realms of social life and thought. The ephemerality of the MONDO space is suggested by the stock of other wax figures kept in the café's storeroom. Sooner or later Bogart and Armstrong are to be replaced by the Beatles, the Rolling Stones and other cultural imports.
210 An unidentified person quoted in Tomas Andersson (1992).
211 Fukuyama (1992), building on his particular readings of Hegel and Kojève, asserts that history has come to an end with the collapse of Communism and the triumph of capitalism and Western liberal democracy. For the antecedents of Fukuyama's much publicized position see Anderson (1992) and Niethammer (1992). Anderson also contains an extended critique of Fukuyama.

AFTER MONTAGE:
OR WHAT?

•

The erosion of the nation-state, national economies and national cultural identities is a very complex and dangerous moment.

STUART HALL
(1991), 25

The more Europe is integrated and the world is globalized, the quicker the dissolution of sedimented practices, routines and traditions proceeds, the all the more national identity is discussed, given a sharper profile and challenged.

JONAS FRYKMAN
(1993), 123

[W]e now have to make sense of a world without stable vantage points; a world in which the observers and the observed are in ceaseless, fluid, and interactive motion, a world where "human ways of life increasingly influence, dominate, parody, translate, and subvert one another."

DEREK GREGORY
(1994), 9[1]

In locally, regionally and nationally variant ways, the hypermodern present characteristic of European commodity societies is an extended moment of danger. Whatever the scale of observation, in each instance there is some conjunction of disturbing instabilities, bewildering contradictions and simmering, if not boiling, tensions.... Everyday encounters with "processes that cross-cut time frames and spatial zones [and one another] in quite uncontrolled [and unpredictable] ways."[2] ... Political deterioration within the member-states and would-be members of the European Union.... Transnational and global economic integration.... National and local economic crisis.... Hypermobile money capital and the quickened quest for new profits, new markets, new labor-cost savings.... Persistently high rates of unemployment, the influx of refugees and other migrant others, and the (re)eruption of racism in various guises.... Cultural conflict and cultural contestation;

struggles over naming, meaning and identity.... Simultaneous and immediate.... Turbulent and frequently vertiginous.... Coupled with the long march of environmental degradation.

How are we to make sense of, to critically examine, these capitalist (hyper)modernities? How are we to intelligibly image(-ine) these complex and volatile intermeshings of political economy and cultural politics? How are we to interrogate and re-present their here and now spatialities as well as their multiple (geographical hi)stories? How are we to show and do from within these circumstances which are at one and the same time locally (regionally, nationally) peculiar and yet largely constituted through material and power-relational interactions with the extra-local (-regional, -national), yet largely constituted through simultaneous entanglement in larger-scale structuring processes that vary in their geographical extent and temporal depth?

The strategy chosen here in meditating upon European capitalist modernities...

Industrial modernity.   Multiple pasts and on-   Hypermodernity.
High modernity.   going presents relayered   High modernity.
Hypermodernity.   with one another.   Industrial modernity.

Brought into constellation with one another.
Like a flash of lightning.

The Stockholm Exhibtion of 1897 was a one-summer-only space at which new mass-market and luxury commodities designed for everday use were offered up as spectacle. As a new (dream)world of goods that was to inspire awe as well as curiosity and desire. As a device for educating the public by way of object lessons, for making some things visible and others invisible – for evoking a sense of progress and national pride while obscuring intensified social tensions. As a promise of an ever better future, of a just-beyond-the-horizon paradise to be achieved through new practices of consumption.

The Stockholm Exhibition of 1930 was a short-lived space where new pure and simple lined consumer goods and housing were spectacularly juxtaposed so as at once to market the idea-logics of high modernity to the Swedish "broad masses" and advertise Swedish goods to the world. So as to promote a break with the past by way of functionalist aesthetics and social engineering. So as to school attendees in the virtues of modern consumption and of the "new [Swedish hu]man"-commitment to solidarity, social responsibility *and* individualism. So as to convince that the present is to be gladly accepted and the future optimistically looked forward to. So as to simultaneously serve the interests of the Social Democratic Party and of corporate capital. So as to once again promise a just-over-the-next-hill paradise to be achieved through new practices of consumption.

The Globe, a more enduring spectacular space of consumption in a place and era where the consumption of spectacles has been reduced to an everyday matter. An ever-present, visible-from-almost-everywhere structure symptomatic of late twentieth-century global forms of capital. A by-product of the global hypermobility of money and the consequent economic competition among large metropolitan complexes for symbolic capital and greater international visibility. A device for marketing Stockholm as an international center of investment opportunities, corporate capital and banking and tourism – and thereby a device to attract economic activity and to generate more local jobs, higher property values and greater retail sales. A "world-class" sports and cultural arena, a spectacular space for one novel commodified entertainment spectacle followed by another novel commodified entertainment spectacle. A space whose history is in keeping with both the increasingly free reign of neoliberalism and market-force advocates in Sweden and the concomitant crisis of the welfare state. A space whose multiple messages are congruent with a number of wider discourses, including that which sings the praises of European economic integration, which claims that European Community (Union) membership will quickly result in greater individual freedom of choice for consumers, in a new consumption-enabled paradise.

The Stockholm exhibitions of 1897 and 1930 and the Globe: three spectacular spaces as much devoted to the consumption of (would-be) hegemonic discourses as to the consumption of new commodity forms. Spectacular spaces of industrial modernity, high modernity and hypermodernity where the quest for greater commodity sales could not (cannot) be untangled from promises of paradise; from discourses designed to maintain, restructure or consolidate domination, to legitimate or extend existing power relations, to ensure or produce a new relative stability within the political and economic domains of concrete everyday life.

Three spectacular spaces at which objects and discourses alike promise much more than can be delivered.

Distinctly different spectacular spaces at distinctly different (dis)junctures of Swedish modernity, promoting distinctly different forms of consumption.

And yet, the always again the same.

The paradise-like next century so heralded at the Stockholm Exhibition of 1897 arrived in the form of economic downturn, falling wages and hard times.

Despite assurances there made that all was going well, that *things* were getting better, the Stockholm Exhibition of 1930 was immediately ensued by a sharp increase in unemployment and the delayed, but full-blown, arrival of the Great Depression.

But three years after its opening, the Globe – THE concretization of Sweden's "Great Consumption Party" – was thoroughly entangled in the country's profound economic crisis. Not only was it necessary to reduce the number of spectacular events. Not only were attendance and revenues down. Globen City KB – the consortium that built the Globe and operated the office, hotel and retailing space of Globen City – had ceased making payments.

> The ceaseless development of new commodities and the tireless search for new markets.
>
> The constant movement of investments from commodity activities with falling rates of return to commodity activities which promise higher rates of return.
>
> Crises of overproduction and overcapacity.
>
> The inevitable obsolescence and death of the more or less recently new, the discarding of the more or less recently purchased.

The always again the same writ large and small.

At the Stockholm Exhibition of 1897 history was rewritten in the form of *Gamla Stockholm* (Old Stockholm), in the form of an illusory trip to a physically palpable past, in the form of purchasable objects and diversions. This rewritten history was at one and the same time to further construct national pride, further fabricate a national heritage and further invent a national community. It was to underline the exhibitionary rhetoric of national and technological progress while providing an escape from the everyday conflicts and cares of the present.

At the Stockholm Exhibition of 1930 history was rewritten in the form of the *Svea Rike* pavilion, in the form of a *Biblia Pauperum*, in the form of a pedagogic "picture book" exhibit that was once again to tell tales of progress. This was history rewritten so as to extol the collective accomplishments of the Swedish "people" (or at least its male "Nordic" types) and to trumpet the triumphs of the State and corporate capital. So as to diminish the unmodern rural and elevate the modern urban. So as to obscure class conflict, political struggle and cultural contestation. So as to "illustrate how our country has achieved the indisputably high [international] standing it now occupies within the economic, social and cultural spheres, and what further development possibilities the future might offer."[3]

At the Globe history is rewritten in the form of large photographic images of "Names we never forget," in the form of black and white pictures of Swedes who made it to the top in "high" or "popular" culture markets during the long or recent past, in the form of a portrait gallery glorifying individual (commodified) success, glorifying women and men who at once represent both the essence of "Swedishness" and (financially rewarding) achievement in the world beyond Sweden's borders. Not least of all, in the form of a visual rhetoric in keeping with the (would-be) hegemonic discourses propagated by free-market interests, in keeping with other efforts on the part of pro-European-Community membership forces to reconfigure Swedish history in term's of the country's long-term economic, political and cultural involvements with the rest of Europe.

The Stockholm exhibitions of 1897 and 1930 and the Globe: three spectacular spaces at which efforts were made to counteract actual or potential dissatisfaction and disenchantment with the here and now
by rewiting history,
by attempting to remythologize the national past,
by attempting to reenchant a once-upon-a-time then and there,
by attempting to reinvent collective memory.

The Stockholm Exhibition of 1897 – consumption as a national project of industrial modernization: "And not one among all the hundred-thousands who surged through the exhibition was completely unmoved by the sense of national solidarity. Everyone understood that this was common property, a common pride; everyone met together in a happy awareness: My people have been capable of this; these glories of labor are mine; so great are the resources of my country; so rich is the nature of my country; so strong and so gifted is my nation."[4]

The Stockholm Exhibition of 1930 – consumption as a national project of social(ly engineered [or high]) modernization: "Iron, wood and waterpower, money, commercial houses and banks are great things, but still greater is the firm resolution of everybody to stand shoulder by shoulder in order to lift our country even higher.... [I]f *Svea Rike* is to reach the goal which is the meaning of its entire history – to rise higher and higher in the world's respect – it depends on every man and woman in the country. Fellow countrymen! Brothers! Sisters! The creation of *Svea Rike*'s future depends on YOU [becoming and acting like a modern citizen]!"[5]

The Globe – consumption as a project of hypermodernization, as simultaneously a local project with a global strategy and a global project given local expression: "[N]othing like the Globe has ever been built anywhere in the world."[6] ... "[A] home of possibilities, a house for everybody; yes, a real all-activity house."[7]

Where in the world am I, are we?

Who in the world am I, are we?

Would-be hegemonic discourses reworked.

The Stockholm Exhibition of 1897: an outburst of grab-anything-you-can-get-your-hands-on looting. Multiply motivated. Different meanings appropriated or brought into question by people with different class, gender, occupational, generational and regional identities.

The Stockholm Exhibition of 1930: a refusal to immediately accept pure and simple lines and their associated future lines of vision. The refusal of pure and simple lined furniture, household goods and housing. Compounded by mocking renamings of the exhibition area and its architecture, by the tee-hees of those who saw through the objectification of the subject and the subjectification of the object inherent in the exhibition's conjoining of consumption and social engineering.

The Globe: passageway graffiti – "The Globe is watching you!" – elicited by a sense of repeated intrusion and domination.... A lexicon of punishing praise–abuse nicknames. The "dick head" bringing down of a politician – a Social Democrat turned market-force advocate – pivotal to the Globe's development; simultaneously an expression of the widespread grassroots dissatisfaction with those no-longer-in-contact, out-of-reach power holders self-isolated at the top of the Social Democratic hierarchy.

The strategy chosen here in meditating upon European capitalist modernities, in attempting to examine critically the extended moment of danger characteristic of hypermodern European commodity societies, in attempting to interrogate and re-present their current spatialities and multiple (geographical hi)stories, has been to devise a montage of the present, a history of the present in montage form. A historical montage – a set of tension-riddled images – through which the industrial-modern and the high-modern are made to speak to the hypermodern present, a historical montage through which ruptured moments of the past acquire significance because of their relationship to the present, a historical montage of radically heterogeneous fragments – of often seemingly inconsequential details – which brings the then and there into tension-filled constellation with the here and now. Stockholm's three spectacular spaces of consumption – and the century of Swedish capitalist modernity to which they are so central – have not served as the empirical foci of meditation because they closely replicate spaces and circumstances elsewhere, because their congruity with related phenomena in other parts of Europe is so great they somehow transcend the geographical and historical distinctiveness of national capitalisms, political circumstances and forms of collective consciousness. On the contrary, they have served as the pivot of my meditation because they constitute an extreme case of European capitalist modernity, because Swedish industrial modernity eventually resulted in the world's highest level of capital concentration (in large commodity-promoting corporations), because Sweden's high (social engineered) modernity came to permeate everyday life more – and to last longer – than in any other European commodity society, because the dissolution of high modernity into hypermodernity has therefore proved especially unsettling in Sweden, *because it is the extreme(s) which can prove most revealing of the central wherever pronounced variety exists around a core terrain of intersections and commonalities, around a deep core of shared interactions, interdependencies and influences.*

The constellations presented above are among the more striking – and obvious – configurations of *recognition* that have emerged from my own encounter with Swedish modernity. But, if at all successful, the radically heterogeneous fragments of each of the previous three chapters, as well as the juxtaposition of each of those chapters with one another, has opened you the reader to other (more or less related?) flashes of understanding, to the possibility of other (more or less related?) mutual illuminations, to other (more or less related?) *recognitions* of European modernity and the present extended moment of danger.

───

As many questions raised as answered.

What are the related but different ways in which industrial modernity and high modernity speak to the precarious hypermodern present elsewhere in Europe?

What are the distant, not so distant and very recent (geographical hi)stories to be brought into tension-filled constellation with the present moment in other European commodity societies?

What, in other European instances, have been the spatially enmeshed, would-be hegemonic discourses through which the State has marketed its legitimacy while corporations and other capitalist enterprises have simultaneously legitimated their marketing (of the new)?

What other montages of the present are to be devised, what other practices of heretical empiricism are to be brought to bear, so as to enable some re*cognition* of the mutual entanglements of modern politics, economy and culture?

What other histories of the present are to be devised so as to make visible the multiple pasts layered within the present moment? So as to startle the reader–observer out of the (dream)world of commodity-society modernity. So as to trigger an awareness of the ever-again-the-same qualities of modern commodity forms and their associated would-be hegemonic discourses, and thereby, so as to render the possibility of sensibilities that are never again the same.

What political actions, if any, are to result from the re*cognition* of European modernities that at once possess shared characteristics – as well as complex interdependencies and interactions past and present – and yet are nationally (and regionally or locally) distinctive? Or from the re*cognition* of the multiple modernities that have existed in any one country and that residually persist in the volatile current moment of fast-capitalism and hypermodernity?

What interrogations and re-presentations are appropriate to other European instances?

What are the tension-riddled images to be evoked?

What is to be shown?

What is to be done?

Every moment of danger is also a moment of opportunity to be seized upon.

Or what?

NOTES

1. Gregory is here quoting Clifford (1986), 22.
2. George E. Marcus (1992), 326.
3. Nordström (1930a), 3.
4. Ellen Key, as quoted in Björck (1947), 360.
5. Nordström (1930a), 9.
6. AB Stockholm Globe Arena (1989), 1.
7. Stefan Holmgren (first President of AB Stockholm Globe Arena) as quoted in Torbjörn Andersson *et al.* (1991), 4.

BIBLIOGRAPHY

•

ARCHIVAL SOURCES

ACM   Archives held at the Stockholm City Museum (*Stockholms Stadsmuseum*). Unless otherwise indicated, refers to *Uppteckningar om lika miljöer, arbetsförhållande och företeelse i Stockholms stad*, (an extensive collection of oral recollections regarding the late nineteenth century, gathered during the 1930s and 1940s).

SMA   Photographic archive of the Swedish Museum of Architecture (*Arkitekturmuseet*).

SU-1930   Archive on the 1930 Stockholm Exhibition held at the Stockholm Municipal Archives (*Stockholms stadsarkiv*).

ETHNOGRAPHIC SOURCES

EMSU   Ethnographic materials collected by students at the Institute of Ethnology at the Nordic Museum and Stockholm University. I am most indebted to Barbro Klein for arranging the collection of these materials.

NEWSPAPER AND JOURNAL SOURCES

Aftonbladet, Stockholm
Allt för Alla, Stockholm
Arbetarbladet, Stockholm
Arbetaren, Stockholm
Dagens Nyheter, Stockholm
Expressen, Stockholm
Faderneslandet, Stockholm
Folket, Stockholm
Folkets Dagblad, Stockholm
Folkets Dagblad Politiken, Stockholm
Göteborgs Handelstidning, Göteborg
Göteborgs Posten, Göteborg
Göteborgs Tidning, Göteborg
Hallands Tidning, Halmstad

Höganäs Tidning, Höganas
Hudiksvallposten, Hudiksvall
Hudikvalls Tidningen, Hudiksvall
Idun, Stockholm
Järnhandlaren, Stockholm
Karlsborgs Tidning, Karlsborg
Kaspar, Stockholm
Katrineholms Kuriren, Katrineholm
Kvällsposten, Göteborg
Lysekils Posten, Lysekil
National Socialisten, Stockholm
Nerikes Tidningen, Örebro
Nordens Expositionstidning, Stockholm
Norrlandsposten, Stockholm
Nya Dagligt Allehanda, Stockholm
Ny Dag, Stockholm
Öbrero-Kuriren, Örebro
Östgöten, Norrköping
Östergötlands Dagblad, Norrköping
Säters Tidning, Säter
Skånska Social-Demokraten, Malmö
Skaratidningen, Skara
Smålands Posten Växjö, Växjö
Social Demokraten, Stockholm
Söndags-Nisse, Stockholm
Söndags-Nisse Strix, Stockholm
Stockholms Dagblad, Stockholm
Stockholms Extrablad, Stockholm
Stockholms Tidningen, Stockholm
Sydsvenska Dagbladet – Snällposten, Malmö
Svensk Reklam, Stockholm
Svensk Skräddertidning, Stockholm
Svenska Dagladet, Stockholm
Svenska Morgonbladet, Stockholm
Västerbottens Kuriren, Umeå
Veckans Affärer, Stockholm

REFERENCES

AB Stockholm Globe Arena (1989) *The Globe Project*, Stockholm: Editech AB.
AB Stockholm Globe Arena (1990) *The Globe – Success in the Round*, Stockholm: Brundin & Sommar.
AB Stockholm Globe Arena (1992) *1991 Annual Report*, Laholm: AB Ruter Press.
Adamson, Walter, L. (1993) *Avant-Garde Florence: From Modernism to Fascism*, Cambridge, Mass.: Harvard University Press.
Aggar, Ben (1989) *Fast Capitalism: A Critical Theory of Significance*, Urbana: University of Illinois Press.
Agnew, Jean-Christophe (1986) *Worlds Apart: The Market and the Theater in Anglo–American Thought*, Cambridge: Cambridge University Press.
Ahlberg, Gösta (1958) *Stockholms befolkningsutveckling efter 1850*, Stockholm: Almqvist & Wiksell.
Ahlmann, H. W:son, Ekstedt, I, Jonsson, G. and William-Olsson, W. (1934) *Stockholms inre differentiering*, Stockholm: Stadskollegiets utlåtande och memorial, bihang nr 51.
Allwood, John (1977) *The Great Exhibitions*, London: Cassell & Collier.
Ambjörnsson, Ronny (1991a) "En skön ny värld – Om Ellen Keys visioner och en senare tids verklighet," *Fataburen*, 260–78.
Ambjörnsson, Ronny (1991b) "Tänk om påven säger att katolicism är ett misstag," *Dagens Nyheter*, December 21.
Anderson, Benedict (1983) *Imagined Communities: Reflections on the Origin and Spread of Nationalism*, London: Verso.

Anderson, Perry (1990) "A culture in contraflow – II," *New Left Review*, 182: 85–137.
Anderson, Perry (1992) *A Zone of Engagement*, London: Verso.
Andersson, Dan and Edin, P.O. (1992) "'Vi flyr in i EG,'" *Dagens Nyheter*, June 17.
Andersson, Georg (1992) "Vad är humant?" *Dagens Nyheter*, 18 July.
Andersson, Henrik O. (1980) "Förord," *Funktionalisms genombrott och kris: Svenskt bostadsbyggande 1930–1980*, Stockholm: ArkitekturMuseet.
Andersson, Roger (1991) *Internationalization, Individualization, and Senses of Community*, Stockholm: FArådet, rapport nr 17.
Andersson, Tomas (1992) "Ny jätte i krogfamiljen," Dagens Nyheter, July 4.
Andersson, Torbjörn, Bergquist, Inga, Fleetwood, Anna Maria, Grahn, Wera, Mattsson, Katarina and Olsson, Anna (1991): "Kulturens roll i Globenprojektet," Stockholm: Institute of Ethnology at the Nordic Museum and Stockholm University, unpublished report.
Andrée, S.A. (1897) *Konsten att studera utställningar*, 2nd edition, Stockholm.
Anesäter, Stig (1989) "Globen och marknaden," *Västerbottens Folkblad*, April 12.
Arrighi, Giovanni (1991) "World income inequalities and the future of socialism," *New Left Review*, 189: 39–65.
Asplund, Gunnar, Gahn, Wolter, Markelius, Sven, Paulsson, Gregor, Sundahl, Eskil and Åhren, Uno (1931) *acceptera*, Stockholm: Bokförlagsaktiebolaget Tiden.
Åström, Sverker (1992) "EG:s försvarssamarbete gagnar Sverige," *Svenska Dagbladet*, August 9.
Bagge, Gösta, Lundberg, Erik and Svennilson, Ingvar (1933) *Wages in Sweden 1860–1930*, part 1, *Manufacturing and Mining*, London: P.S. King & Son, Ltd.
Bakhtin, Mikhail (1984) *Rabelais and His World*, Bloomington: Indiana University Press.
Barthes, Roland (1979) *The Eiffel Tower and Other Mythologies*, New York, Hill & Wang.
Baudrillard, Jean (1983) "The ecstasy of communication," in H. Foster, (ed.) *The Anti-Aesthetic: Essays on Postmodern Culture*, Seattle: Bay Press, 126–34.
Bendix, Carl L. (1899) "Inträdesreglemente, samtrafik, besökande," in Looström, L. (ed.) *Allmänna Konst- och Industriutställningen i Stockholm 1897. Officiel berättelse*, Stockholm, 109–24.
Benedict, Burton (1983) *The Anthropology of World's Fairs: San Francisco's Panama Pacific International Exposition of 1915*, Berkeley: Scolar Press.
Benjamin, Walter (1972–1989) *Gesammelte Schriften*, Frankfurt am Main: Suhrkamp Verlag 7 volumes.
Benjamin, Walter (1976) *Charles Baudelaire: A Lyric Poet in the Era of High Capitalism*, London: Verso.
Benjamin, Walter (1978) *Reflections: Essays, Aphorisms, Autobiographical Writings*, edited and with an introduction by Peter Demetz, New York and London: Harcourt Brace Jovanovich.
Benjamin, Walter (1979) *One Way Street and Other Writings*, London: New Left Books.
Benjamin, Walter (1982) *Gesammelte Schriften*, vol. 5, *Das Passagen-Werk*, edited by Rolf Tiedemann, Frankfurt am Main: Suhrkamp Verlag.

Bennett, Tony (1988) "The exhibitionary complex," *New Formations*, 4: 73–102.
Berger, Peter L. and Luckmann, Thomas (1967) *The Social Construction of Reality: A Treatise in the Sociology of Knowledge*, New York: Anchor Books.
Bildt, Carl (1992) "Säkerhetspolitiknen självklar i EG-samarbetet," *Svenska Dagbladet*, August 11.
Björck, S. (1947) "Stockholmsutställningen 1897 som tidsspegel," *Ord och Bild*, 56: 353–60.
Björklund, Tom (1967) *Reklamen i svensk marknad 1920–1965 – En ekonomisk-historisk återblick på marknadsförings – och reklamutvecklingen efter första världskriget*, 2 vols., Stockholm: P.A. Norstedt och Söners förlag.
Björkman, Helena (1989) "Spherical buildings," in Claes Dymling (ed.) *Stockholm Globe Arena: A Document on Its Conception and Creation*, Stockholm: Byggförlaget, 97–102.
Blanche, Thore (1897) *Officiel vägvisare öfver Allmänna Konst – och Industriutställningen i Stockholm 1897*, Stockholm.
Bloomfield, Terry (1991) "It's sooner than you think, or where are we in the history of rock music," *New Left Review*, 190: 59–81.
Bondi, Liz (1993) "Locating identity politics," in M. Keith and J. Pile (eds.) *Place and the Politics of Identity*, London: Routledge, 84–101.
Bonnett, Alastair (1989) "Situationism, geography and poststructuralism," *Society and Space*, 7: 131–46.
Bourdieu, Pierre (1991) *Language and Symbolic Power*, Cambridge, Mass.: Harvard University Press.
Boyer, M. Christine (1992) "Cities for sale: merchandising history at South Street Seaport," in Michael Sorkin (ed.) *Variations on a Theme Park: The New Ameican City and the End of Public Space*, New York: The Noonday Press, 181–204.
Britton, S. (1991) "Tourism, capital, and place: towards a critical geography of tourism," *Society and Space*, 9: 451–78.
Broberg, Gunnar and Tydén, Mattias (1991) *Oönskade i folkhemmet – Rashygien och steriliseringen i Sverige*, Stockholm: Gidlunds.
Brunius, Teddy (1989) "Otto G. Carlsund – En efterskrift," in Brunius, Teddy and Moberg, Ulf Thomas (eds.) *Om och Av Otto G. Carlsund*, Stockholm: Cinclus, 181–202.
Bruno, William (1954) *Tegelindustrien i Mälarprovinserna 1815–1950 med särskild hänsyn till Stockholm som marknad*, Stockholm. Geografica – Skrifter från Upsala Universitets Geografiska institutionen, 28.
Buck-Morss, Susan (1986) "The flaneur, the sandwichman and the whore: the politics of loitering," *New German Critique*, 39: 99–140.
Buck-Morss, Susan (1989) *The Dialectics of Seeing: Walter Benjamin and the Arcades Project*, Cambridge, Mass.: The MIT Press.
Bürger, Peter (1984) *Theory of the Avant-Garde*, Minneapolis: University of Minneapolis Press.
Calinescu, Matei (1977) *Faces of Modernity: Avant-Garde, Decadence, Kitsch*, Bloomington: Indiana University Press.

Calinescu, Matei (1978) "Om kitsch," *Jakobs Stege: Coeckelbergs Litterära Tidskrift*, 2/3: 30–54.

Carlberg, Ingrid (1992) "Långt borta och nära i EG," *Dagens Nyheter*, June 26.

Carlsund, Otto G. (1930): "Introduktion" to the catalog *Internationell utställning av post-kubistisk konst 19 Aug.–30 Sept.*, in Teddy Brunius, and Ulf Thomas Moberg (eds.) *Om och Av Otto G. Carlsund*, Stockholm: Cinclus, 77–9.

Celik, Zeynep (1992) *Displaying the Orient: Architecture of Islam at Nineteenth-Century World's Fairs*, Berkeley: University of California Press.

Childs, Marquis W. (1936) *Sweden: The Middle Way*, New York: Penguin Books.

Christensen, Anna (1992) "Den enda vägen leder alltid fel," *Dagens Nyheter*, July 27.

Christofferson, Birger (1961) "Förord," in Birger Christofferson and Thomas von Vegesack (eds.) *Perspektiv på 30-talet*, Stockholm: Wahlström & Widstrand, 7–8.

Clark, T.J. (1984) *The Painting of Modern Life: Paris in the Art of Manet and his Followers*, Princeton: Princeton University Press.

Clifford, James (1986) "Introduction: Partial truths," in James Clifford and George E. Marcus (eds.) *Writing Culture: The Poetics and Politics of Ethnography*, Berkeley: University of California Press, 1–26.

Cornell, Elias (1952) *De stora utställningarnas arkitekturhistoria*, Stockholm: Bokförlaget Natur och Kultur.

Conradson, Birgitta (1989) "Den kulturskapande reklamen: om reklammediernas framväxt efter industrialisms genombrott," *Företagsminnen – Års meddelande från Föreningen Stockholms företagsminnen*, 12–21.

Crawford, Margaret (1992) "The world in a shopping mall," in Michael Sorkin (ed.) *Variations on a Theme Park*, New York: Hill & Wang, 3–30.

Cronon, William (1991) *Nature's Metropolis: Chicago and the Great West*, New York: W.W. Norton & Company.

Daun, Åke (1989) *Svensk mentalitet – ett jämförande perspektiv*, Stockholm: Raben & Sjögren.

Daun, Åke (1992) *Den europeiska identiteten – bidrag till samtal om Sveriges framtid*, Stockholm: Raben & Sjögren.

Davison, Graeme (1982) "Exhibitions," *Australian Cultural History*, 2: 5–21.

Debord, Guy (1971) *La Société du Spectacle*, Paris: Champ Libre.

Debord, Guy (1983) *The Society of the Spectacle*, Detroit: Red & Black.

Debord, Guy (1990) *Comments on the Society of the Spectacle*, London: Verso.

de Certeau, Michel (1984) *The Practice of Everyday Life*, Berkeley: University of California Press.

Denning, M. (1990) "The end of mass culture," *International Labor and Working Class History*, 37: 4–18.

Doordan, Dennis P. (1988) *Building Modern Italy: Italian Architecture, 1914–1936*, New York: Princeton Architectural Press.

Dunbar, Gary S. (1974) "Élisée Reclus and the great globe," *Scottish Geographical Magazine*, 90: 57–66.

Dunbar, Gary S. (1978) *Élisée Reclus: Historian of Nature*, Hampden, Conn.: Archon Books.

Duncan, J.S. (1990) *The City as Text: The Politics of Landscape Interpretation in the Kandyan Kingdom*, Cambridge: Cambridge University Press.
Dymling, Claes (ed.) (1989) *Stockholm Globe Arena: A Document on Its Conception and Creation*, Stockholm: Byggförlaget.
Edwards, Folke (1980) "Den funktionella rakkniven," *1930/80 – Arkitektur Form Konst*, Stockholm: Kulturhuset, 20–34.
Ehrensvärd, U. (1972) *Gamla vykort*, Stockholm: Bonniers.
Ekbom, Torsten (1991) "Kabaré födde modernism," *Dagens Nyheter*, 11 November 1991.
Eksteins, Modris (1990) *Rites of Spring: The Great War and the Birth of the Modern Age*, New York: Anchor Books.
Ekström, Anders (1989) "International exhibitions and the struggle for cultural hegemony," *Uppsala Newsletter: History of Science*, 12: 6–7.
Ekström, Anders (1991a) "Industriexpositionen- en läroanstalt," *Tvärsnitt*, 1: 102–9.
Ekström, Anders (1991b) "Stockholmsutställningen 1897," *Fataburen*, 97–129.
Enders, Georg and Nils-Georg (1930) *Det är vår sommarmelodi*, Stockholm: Nils-Georgs Musikförlag.
Eriksson, Eva (1990) *Den Moderna Stadens Födelse – Svensk Arkitektur 1890–1920*, Stockholm: Ordfronts Förlag.
Etlin, Richard (1991) *Modernism in Italian Architecture, 1890–1940*. Cambridge, Mass.: The MIT Press.
Ewen, Stuart (1988) *All Consuming Images: The Politics of Style in Contemporary Culture*, New York: Basic Books.
Falskhet (1901) Stockholm.
Featherstone, Mike (1991) *Consumer Culture and Postmodernism*, London: Sage Publications.
Feldt, Kjell-Olof (1991) *Alla dessa dagar ... I regeringen 1982–1990*, Stockholm: Norstedts.
Folcker, E.G. (1897) "Utställningen: En öfverblick," *Ord och Bild*, 6: 272–80, 371–80.
Folcker, E.G. (1899) "Gamla Stockholm," in Looström, L. (ed.) *Allmänna Konst – och Industriutställningen. Officiel berättelse*, Stockholm.
Forty, A. (1986) *Objects of Desire: Design and Society, 1750–1980*, London: Thames & Hudson/Cameron.
Foster, Hal (1985) *Recodings: Art, Spectacle, Cultural Politics*, Seattle: Bay Press.
Foucault, Michel (1972) *The Archeology of Knowledge*, New York: Pantheon.
Foucault, Michel (1975) "Film and popular memory: an interview," *Radical Philosophy*, 11: 24–9.
Foucault, Michel (1978) *The History of Sexuality, I: An Introduction*, New York: Pantheon.
Frampton, Kenneth (1985) "Stockholm 1930: Asplund and the legacy of Funkis," in Claes Caldenby and Olof Hultin (eds.) *Asplund*, Stockholm: Arkitektur Förlag, 35–9.
Frampton, Kenneth (1992) *Modern Architecture – A Critical History*, 3rd edition, London: Thames & Hudson.

Friggebo, Birgit and Westerberg, Bengt (1992) "En human flyktningspolitik," *Dagens Nyheter*, July 11.
Frisby, David (1986) *Fragments of Modernity: Theories of Modernity in the Work of Simmel, Kracauer and Benjamin*, Cambridge, Mass.: The MIT Press.
Frykman, Jonas (1981) "Pure and rational – the hygienic vision: a study of cultural transformation in the 1930s," *Etnologia Scandinavica*, 36–63.
Frykman, Jonas (1993) "Nationella ord och handlingar," in Billy Ehn, Jonas Frykman and Orvar Löfgren, *Försvenskningen av Sverige – det nationellas förvandlingar*, Stockholm: Natur och Kultur, 119–201.
Frykman, Jonas and Löfgren, Orvar (1979) *Den kultiverade människan*, Lund: Liber Läromedel.
Frykman, Jonas and Löfgren, Orvar (1985) "På väg- bilder av kultur och klass," in Jonas Frykman, Orvar Löfgren, *et al.* (eds) *Modärna tider – Vision och vardag i folkhemmet*, Malmö: Liber Förlag, 20–139.
Frykman, Jonas and Löfgren, Orvar (1987) *Culture Builders: A Historical Anthropology of Middle-Class Life*, New Brunswick: Rutgers University Press.
Fukuyama, Francis (1992) *The End of History and the Last Man*, New York: Avon.
Fuller, Mia (1988) "Building power: Italy's colonial architecture and urbanism," *Cultural Anthropology*, 3: 455–87.
Fürth, Thomas (1979) *De arbetslösa och 1930-talskrisen – En kollektivbiografi över hjälpsökande arbetslösa i Stockholm 1928–1936*, Stockholm: Monografier utgivna av Stockholms kommun, 40.
Galison, Peter (1990) "Aufbau/Bauhaus: logical positivism and architectural modernism," *Critical Inquiry*, 16: 709–52.
Gargas, Jacqueline (ed.) (1981) "From futurism to rationalism: the origins of modern Italian architecture," *Architectural Design*, 51, 1/2: 1–80.
Gaunt, David (1983) *Familjeliv i Norden*, Stockholm: Gidlunds.
Gaunt, David and Löfgren, Orvar (1984) *Myter of Svensken*, Stockholm: Liber Förlag.
Geijerstam, Gustaf af (1894) *Anteckningar om arbetarförhållanden i Stockholm*, Stockholm.
"Georgie" (1898) "Tempel i tiden," *Skånska Dagbladet*, February 23.
Giddens, Anthony (1991) *Modernity and Self-Identity: Self and Society in the Late Modern Age*, Stanford Conn.: Stanford University Press.
Göransson, Eva (1989) "Fy bubblan för Makten," *Expressen*, February 26.
Gramsci, Antonio (1971) *Selections from the Prison Notebooks*, New York: International Publishers.
Greenhalgh, Paul (1988) *Ephemeral Vistas: The Expositions Universelles, Great Exhibitions and World's Fairs, 1851–1939*, Manchester: Manchester University Press.
Gregory, Derek (1991) "Interventions in the historical geography of modernity: social theory, spatiality and the politics of representation," *Geografiska Annaler*, 73B: 17–44.
Gregory, Derek (1994) *Geographical Imaginations*, Oxford: Blackwell.
Grossberg, Lawrence (1986) "On postmodernism and articulation: an interview with Stuart Hall," *Journal of Communication Inquiry*, 10: 53–5.
Grossberg, Lawrence (1992) *We Gotta Get Out of This Place: Popular Conservatism and*

Postmodern Culture, New York and London: Routledge.

Gustafson, Uno (1976) *Studier rörande Stockholms sociala, ekonomiska och demografiska struktur, 1860–1910*, Monografier utgivna av Stockholms kommunalförvaltning, 37.

Hall, Stuart (1981) "Cultural studies in two paradigms," in Tony Bennett *et al.* (eds.) *Culture, Ideology and Social Process*, London: Batsford, 19–37.

Hall, Stuart (1988) "The toad in the garden: Thatcherism among the theorists," in Cary Nelson and Lawrence Grossberg (eds.) *Marxism and the Interpretation of Culture*, Urbana: University of Illinois Press, 35–7.

Hall, Stuart (1989) "Cultural identity and cinematic representation," *Framework*, 36: 69–70.

Hall, Stuart (1991) "The local and the global: globalization and ethnicity," in Anthony D. King (ed.) *Culture, Globalization and the World-System: Contemporary Conditions for the Representation of Identity*, Binghamton: Department of Art and Art History, SUNY Binghamton, 19–39.

Hall, Thomas (1989) "En rymdålderns Hattstuga," *Dagens Nyheter*, February 19.

Hamilton, Carl and Rolander, Dag (1992) "Inflationen inget problem," *Dagens Nyheter*, 19 December.

Hammarström, Ingrid (1970) *Stockholm i svensk ekonomi 1850–1914*, Stockholm: Monografier utgivna av Stockholms kommunalförvaltning 22: 2.

Haraway, Donna (1991) *Simians, Cyborgs and Women: The Reinvention of Nature*, New York: Routledge.

Harvey, David (1988) "Voodoo Cities," *New Statesman and Society*, 30 September.

Harvey, David (1989a) "From managerialism to entrepreneurialism: the transformation of urban governance in late capitalism," *Geografiska Annaler* 71B: 3–17.

Harvey, David (1989b) *The Condition of Postmodernity: An Enquiry into the Origins of Cultural Change*, Oxford: Basil Blackwell.

Harvey, David (1993) "From space to place and back again: reflections on the condition of postmodernity," in Jon Bird, Barry Curtis, Tim Putnam, George Robertson and Lisa Tickner (eds.) *Mapping the Futures: Local Cultures, Global Change*, London: Routledge, 3–29.

Hasselgren, A. (1897) *Utställningen i Stockholm 1897*, Stockholm.

Hayden, Dolores (1981) *The Grand Domestic Revolution: A History of Feminist Designs for American Homes, Neighborhoods and Cities*, Cambridge, Mass.: The MIT Press.

Henderson, Jeffrey and Castells, Manuel (eds.) (1987) *Global Restructuring and Territorial Development*, London: Sage Publications.

Hirdman, Yvonne (1983) *Matfrågan – Mat som mål och medel, Stockholm 1870–1920*, Stockholm: Rabén och Sjögren.

Hirdman, Yvonne (1990) *Att lägga livet till rätta-Studier i svensk folkhemspolitik*, Stockholm: Carlssons.

Hobsbawm, Eric J. (1990) *Nations and Nationalism since 1780*, Cambridge: Cambridge University Press.

Hoggett, P. (1992) "A place for experience: a psychoanalytic perspective on boundary, identity, and culture," *Society and Space* 10: 345–56.

Holm, Hans Axel (1992) "En kul kula," *Dagens Nyheter*, November 29.

Hughes, Robert (1991) *The Shock of the New*, revised edition, New York: Alfred A. Knopf.

Hunter, Ian (1988) "Setting limits to culture," *New Formations*, 4: 103–123.

Huyssen, Andreas (1986) *After the Great Divide: Modernism, Mass Culture, Postmodernism*, Bloomington: Indiana University Press.

Idestam-Almquist, B. (1948) "Skuggspel och Nick Carter, filmens föregångare," in G. Berg et al. (eds.) *Det Glada Sverige – Våra fester och högtider genom tiderna*, Stockholm: Natur och Kultur, 3, 2,231–2,264.

Illustrerade familjetidskriften för svenska hem (1897) *Allmänna konst – och industriutställningen i Stockholm 1897 – gratisbilaga*, magazine supplement.

Illustrerad handbok under Allmänna Konst – och Industriutställningen i Stockholm 1897 (1897) Stockholm.

Internationell utställning av post-kubistist konst August 19–September 30 (1930) reproduced in Brunius, Teddy and Moberg, Ulf Thomas, (eds.) *Om och Av Otto G. Carlsund*, Stockholm: Cinclus, 75–120.

Jakobsson, Svante (1977) "Rotemansinstitutionen i Stockholm för folkbokföring och mantalskrivning m.m." Uppsala: mimeographed.

Jenkins, David (1969) *Sweden: The Progress Machine*, London: Robert Hale.

Johanneson, L. (1978) *Den mass producerade bilden – ur bildindustrialisms historia*, Stockholm: AWE/Gebers.

Johansson, Gotthard (1929) "Romantik och realism vid Djurgårdsbrunnsviken," *Svenska Dagbladet*, 27 October.

Johansson, Gotthard (1930) "Svea rike i totalistisk projektion," *Svenska Dagbladet*, May 27.

Johansson, Gotthard (1942) "Trettiotalets Stockholm," *Samfundet S:t Eriks Årsbok*, 165–235.

Johansson, Gothard (1955) "Lever funktionalism?," *a5- Meningsblad før unge arkitekter*, 8 (2nd series), June, 64.

Johnson, Eyvind (1948) *Krilon – en roman om det sannolika*, Stockholm: Albert Bonniers Förlag.

Johnson, Eyvind, (1961) "En betraktares förutsättningar," "På jakt efter källor," "Mellan idyll och hemskhet,""Ett decennium utan slut," in Birger Christofferson and Thomas von Vegesack (eds.) *Perspektiv på 30-talet*, Stockholm: Wahlström & Widstrand, 9–39.

Karling, Torsten (1930) "Reklamen på Stockholmsutställningen," *Affärsekonomi*, 9: 399–419.

Karlström, Ulla (1989) "Managing a unique project," in Claes Dymling (ed.) *Stockholm Globe Arena: A Document on Its Conception and Creation*, Stockholm: Byggförlaget, 14–21.

Katalog över bostadsavdelningen – Stockholmsutställningen 1930 (1930), Stockholm: Bröderna Lagerström.

Kellner, Douglas (1989) *Jean Baudrillard: From Marxism to Postmodernism and Beyond*, Stanford: Stanford University Press.

Key-Åberg, Karl (1897a) *En undersökning af arbetares bostadsförhållanden i Stockholm*, Stockholm.
Key-Åberg, Karl (1897b) "Stockholms industri, handel och sjöfart," in E. Dahlgren (ed.) *Stockholm – Sveriges hufvudstad skildrad med anledning af Allmänna Konst och Industriutställningen 1897*, Stockholm, 3: 1–59.
King, Anthony D. (1992) "Identity and difference: the internationalization of capital and the globalization of culture," in Paul L. Knox (ed.) *The Restless Urban Landscape*, Englewood Cliffs, N.J.: Prentice-Hall, 83–100.
Klein, Barbro (1994) "Kraftzoner och symboliska resor: rituella transformationer av Stockholms gator och torg." Unpublished manuscript to appear in altered form in Barbro Klein (ed.) (1995) *När gatan blir vår: Ritualer på offentliga platser*, Stockholm: Carlssens.
af Klintberg, Bengt (1988) "Etnocentriska sägner," in Åke Daun and Billy Ehn (eds.) *Blandsverige – Om kulturskillnader och kulturmöten*, Stockholm: Carlssons, 171–81.
Knight, Richard V. and Gappert, Gary (eds.) (1989) *Cities in a Global Society*, London: Sage Publications.
Knox, Paul L. (1992) "Capital, material culture and socio-spatial differentiation," in Paul L. Knox (ed.) *The Restless Urban Landscape*, Englewood Cliffs, N.J.: Prentice-Hall, 1–34.
Kommittén för utredning av allmän svensk utställning av konstindustri, konsthantverk och hemslöjd i Stockholm (1928a) *Till konungen*, January 28, 1928, petition to the Crown.
Kommittén för utredning av allmän svensk utställning av konstindustri, konsthantverk och hemslöjd i Stockholm (1928b) *Till konungen*, April 27, 1928, petition to the Crown.
Kylhammar, M. (1990) *Den okände Sten Selander – En borgerlig intellektuell*, Stockholm: Akademeja.
L., H. (1930) "Funktionalism och den gifta fabriksarbetarskan," *Morgonbris*, 26, 7 (July): 2–3.
Laclau, Ernesto (1988) "Metaphor and social antagonisms," in Cary Nelson and Lawrence Grossberg (eds.) *Marxism and the Interpretation of Culture*, Urbana: University of Illinois Press, 249–57.
Laclau, Ernesto and Mouffe, Chantal (1985) *Hegemony and Socialist Strategy: Towards a Radical Democratic Politics*, London: New Left Books.
Larsson, Yngve (1967), *På marsch mot demokratin – från hundragradig skala till allmän rösträtt*, monografier utgivna av Stockholms kommunalförvaltning, 22, 4.
Larsson, Yngve (1977) *Mitt liv i stadshuset*, 2 vols., Stockholm: monografier utgivna av Stockholms kommunalförvaltning, 22, IV:2, 22, IV:3.
Levaco, Ronald (1974) *Kuleshov on Film: Writings by Lev Kuleshov*, Berkeley: University of California Press.
Lewis, Russell (1983) "Everything under One Roof: World's Fairs and Department Stores in Paris and Chicago," *Chicago History*, 12, 3: 28–47.
Ley, David and Olds, K. (1988) "Landscape as spectacle: world's fairs and the culture of heroic consumption," *Society and Space*, 6: 191–212.
Liljefors, Åke and Bergquist, Sven (1992) "Statistiken ger osann bild av svensk industri," *Svenska Dagbladet*, June 21.

Lind, Ingela (1991) "Folkhemmets präktiga värden," *Dagens Nyheter*, October 10.
Lindorm, E. (1934) *Oscar II och hans tid – En bokfilm*, Stockholm: Wahlström & Widstrand.
Lindqvist, Sven (1992) "Ska kyrkan passa lilla Lisa?," *Dagens Nyheter*, September 7.
Loader, Christina (1983) *Russian Constructivism*, New Haven, Conn.: Yale University Press.
Lodin, Sven-Olof (1992) "Industrins fördelar överdrivs," *Dagens Nyheter*, June 22.
Löfgren, Mikael (1991) "Drömspelet som dröm," *Dagens Nyheter*, March 23.
Löfgren, Orvar (1989a) "The nationalization of culture," *Ethnologia Europaea*, 19: 5–24.
Löfgren, Orvar (1989b) "Landscapes and mindscapes," *Folk*, 31: 183–208.
Löfgren, Orvar (1989c) "Learning to remember and learning to forget: class and memory in modern Sweden," *Beiträge zur Volkskunde in Niedersachsen*, 5: 145–61.
Löfgren, Orvar (1990a) "Being a good Swede: national identity as a cultural battleground," paper presented at a workshop on *The National Experience*, Bjärsjölagård.
Löfgren, Orvar (1990b) "Consuming interests," *Culture and History*, 7: 7–36.
Löfgren, Orvar (1990c) "Tingen och tidsandan," in A. Arvidsson, *et al.* (eds.) *Människor och föremål – Etnologer om materiell kultur*, Stockholm: Carlssons, 187–208.
Löfgren, Orvar (1991a) "Att nationalisera moderniteten," in Anders Linde-Laursen and Jan Olof Nilsson (eds.) *Nationella identiteter i Norden – Ett fullbordat projekt?*, Eskilstuna: Nordiska rådet, 101–15.
Löfgren, Orvar (1991b) "Nationalizing modernity," paper presented at a symposium on Comparative Modernities, Lund.
Löfgren, Orvar (1992) "Swedish modern: konsten att nationalisera konsumtion och estetik," *Kulturstudier*, 17: 159–80.
Lo-Johansson, Ivar (1957) *Författaren*, Stockholm: Albert Bonniers Förlag.
Lo-Johansson, Ivar (1974) *Lastbara berättelser*, Stockholm: Bonniers.
Lo-Johansson, Ivar (1979) *Asfalt*, Stockholm: Bonniers.
Lööw, Helene (1990) *Hakkorset och Wasakärven – En studie av nationalsocialism i Sverige 1924–1950*, Göteborg: Avhandlingar från Historiska institutionen i Göteborg, nr. 2.
Looström, Ludv. (1899a). "Utställningens förhistoria," in L. Looström (ed.) *Allmänna Konst – och Industriutställningen i Stockholm 1897, Officiel berättelse*, Stockholm, 1–18.
Looström, Ludv. (1899b) "Den femtonde Maj. Utställningens högtidliga invigning," in Ludv. Looström (ed.) *Allmänna Konst – och Industriutställningen i Stockholm 1897, Officiel berättelse*, Stockholm, 101–8.
Looström, Ludv. (1899c) "Utställningens organisation," in Ludv. Looström (ed.) *Allmänna Konst – och Industriutställningen i Stockholm, Officiel berättelse*, Stockholm, 19–54.
Looström, Ludv. (1900) "Juryn och prisutdelningen. Utställningens afslutande," in Ludv. Looström (ed.) *Allmämma Konst– och Industriutställningen, Officiel berättelse*, Stockholm, 1,059–72.

Luke, Timothy W. (1989) *Screens of Power: Ideology, Domination, and Resistance in Informational Society*, Urbana: University of Illinois Press.

Lundgren, Bo (1992) "Inriktningen är att sänka skatteuttaget," *Svenska Dagbladet*, July 26.

Lundgren, M. (1990). *På tröskeln till en ny tid- en analys av den allmänna konst – och industriutställningen i Stockholm 1897*, Uppsala: mimeographed.

Lundkvist, Artur (1961) "Parenteser mellan katastrofer," in Birger Christofferson and Thomas von Vegesack, (eds.) *Perspektiv på 30-talet*, Stockholm: Wahlström & Widstrand, 7–8.

Lundqvist, Åke (1992) "Nu ska Sahlén städa upp i finansruinerna," *Dagens Nyheter*, November 7.

Lyttkens, Lorentz (1991) *Uppbrottet från lagom – En essä om hur Sverige motvilligt tar sig in i framtiden*, Stockholm: Akademeja.

Mannheimer, Otto (1992) "Pizza brer ut sig norröver," *Dagens Nyheter*, July 4.

Marcus, George E. and Fischer, Michael M.J. (1986) *Anthropology as Cultural Critique: An Experimental Moment in the Human Sciences*, University of Chicago Press: Chicago.

Marcus, George E. (1992) "Past, present and emergent identities: requirements for ethnographies of late twentieth-century modernity worldwide," in Scott Lash and Jonathan Friedman (eds.) *Modernity and Identity*, Oxford: Blackwell, 309–30.

Marcus, Greil (1989) *Lipstick Traces: A Secret History of the Twentieth Century*. Cambridge, Mass.: Harvard University Press.

Massey, Doreen (1993) "Power-geometry and a progressive sense of place." In Jon Bird, Barry Curtis, Tim Putnam, George Robertson and Lisa Tickner (eds.) *Mapping the Futures: Local Cultures, Global Change*, London: Routledge, 59–69.

Matovic, Margareta A. (1984) *Stockholmsäktenskap – Familjebildning och partnerval i Stockholm 1850–1890*, monografier utgivna av Stockholms kommun, 57.

Matrikel: Allmänna Konst- och Industriutställningen (1897) Stockholm.

Meregelli, Marta (1982) *L'Esposizione di Stoccolma del 1930: Architettura Funzionalista e Industrial Design*, Milan: Università degli studi di Milano.

Miller, Daniel (1987) *Material Culture and Mass Consumption*, Oxford: Basil Blackwell.

Miller, Roger (1992) "When architecture betrays ideology: post-modernism versus social democracy in recent building projects in Stockholm, Sweden," unpublished paper.

Mitchell, Timothy (1991) *Colonizing Egypt*, Berkeley: University of California Press.

Molander, H. (1899) "Folktågen," in Looström, Ludv. (ed.) *Allmänna Konst – och Industriutställningen i Stockholm 1897, Officiel berättelse*, Stockholm, 148–52.

Montgomery, G.A. (1939) *The Rise of Modern Industry in Sweden*, London: P.S. King & Son, Ltd.

Morgan, Jane, O'Neill, Christopher and Harré, Rom (1979) *Nicknames: Their Origins and Social Consequences*, London: Routledge and Kegan Paul.

Mossige-Norheim, Randi (1988) "Globen: Storföretagens lekstuga," *Folket i Bild Kulturfront*, 6: 2–3, 12–13.

Mumford, Lewis (1934) *Technics and Civilization*, New York: Harcourt Brace.

Munthe, Gustaf (1948) "Gratisnöjen under 1800-talet," in G. Berg *et al.* (eds.) *Det glada Sverige: Våra fester och högtider genom tiderna*, Stockholm: Natur och Kultur, 3: 1,603–1,641.
Myrdal, Alva and Myrdal, Gunnar (1934) *Kris i befolkningsfrågan*, Stockholm: Bonniers.
Myrdal, Gunnar (1933) *The Cost of Living in Sweden, 1830–1930*, London: P.S. King & Son, Ltd.
N., T. (1897) *Trulls' och Anna Stinas minnen från utställningen 1897*, Stockholm.
Näsström, Gustaf (1930a) "Plejaden: Parkrestaurangen," *Stockholms Dagblad*, September 2.
Näsström, Gustaf (1930b) "Samfärdshallen," *Stockholms Dagblad*, May 5.
Näsström, Gustaf (1961) "Från Funkis till Fritiden," in Birger Christofferson and Thomas von Vegesack (eds.) *Perspektiv på 30-talet*, Stockholm: Wahlström & Widstrand, 40–5.
Naylor, Gillian (1990) "Swedish grace ... or the acceptable face of modernism," in Paul Greenhalgh (ed.) *Modernism in Design*, London: Reaktion Books, 164–239.
Niethammar, Lutz (1992) *Posthistoire*, London: Verso.
Nilsson, Jan Olof (1991) "Modernt, allt för modernt – Speglingar," in Anders Linde-Laursen and Jan Olof Nilsson (eds.) *Nationella identiter i Norden – Ett fullbordadat projekt?*, Eskilstuna: Nordiska rådet, 59–99.
Nordström, Ludvig (1930a) *Svea Rike*, Stockholm: Albert Bonniers Förlag.
Nordström, Ludvig (1930b) "Svea Rike," *Svenskt Turistväsen*, 7, 1: 1–7.
Nylén, Leif (1991) "Funkis för folket," *Dagens Nyheter*, April 28.
Nyström, Bertil (1930) "'Svea Rike' på Stockholmsutställningen," *Sunt Förnuft*, June 3–6.
Officiel huvudkatalog – Stockholmsutställningen 1930 av konstindustri, konsthantverk och hemslöjd (1930), Uppsala: Almqvist & Wiksells.
Ohlsson, Martin A. (1959) "De stora utställningarna," in G. Etzler *et al.* (eds.) *Levande stad – en bok om Stockholm*, Stockholm: Rabén & Sjögren, 185–202.
Olsson, Axel Adolf (1947) *Minnen från Stockholmsutställningen och sommaren år 1897*, document, ACM.
Olsson, Gunnar (1990) *Antipasti*, Göteborg: Bokförlaget Korpen.
Olsson, Gunnar (1991a) "Invisible maps: a prospective," *Geografiska Annaler*, 73B: 85–91.
Olsson, Gunnar (1991b) *Lines of Power/Limits of Language*, Minneapolis: University of Minneapolis Press.
Ordningsföreskrifter och andra upplysningar för utställare och hyresgäster (1930), Stockholm: Bröderna Lagerström.
Östlundh, Håkan (1989) "The globe takes shape," in Claes Dymling (ed.) *The Stockholm Globe Arena: A Document on Its Conception and Creation*, Stockholm: Byggförlaget, 92–6.
Paasi, Anssi (1986): "The institutionalization of regions: a theoretical framework for understanding the emergence of regions and the constitution of regional identity," *Fennia*, 164: 105–46.

Paasi, Anssi (1991) "Deconstructing regions: notes on the scales of spatial life," *Environment and Planning A*, 23: 239–56.
Patton, Phil (1993) "'Sell the cookstore if necessary, but come to the Fair,'" *Smithsonian Magazine*, 24, 3: 38–51.
Paulsson, Gregor (1928) *Stockholmsutställningens program*, Stockholm: Bröderna Lagerström.
Paulsson, Gregor (1930a) "Inledning," *Katalog över bostadsavdelningen – Stockholmsutställningen 1930*, Stockholm: Bröderna Lagerström, 23–4.
Paulsson, Gregor (1930b) "Konstindustrien och den nya tiden," *Officiel huvudkatalog – Stockholmsutställningen 1930 av konstindustri, konsthantverk och hemslöjd*, Uppsala: Almqvist & Wiksells, 35–7.
Paulsson, Gregor (1937) *Redogörelse för Stockholmsutställningen 1930*, Uppsala: mimeographed.
Paulsson, Gregor (1970) *Upplevt*, Stockholm: Natur och Kultur.
Persson, Kristina and Rexed, Knut (1992) "Misstro nationell demokrati," *Dagens Nyheter*, July 4.
Powell, Kenneth (1991): "Modernism divided," in Justin Ageros and Catherine Cooke (eds.) *The Avant-Garde: Russian Architecture in the Twenties*. [Architectural Design Profile no. 93], London: Academy Editions, 6–7.
Pred, Allan (1983) "Structuration and place: on the becoming of sense of place and structure of feeling," *Journal for the Theory of Social Behavior*, 13: 157–86.
Pred, Allan (1990a) *Lost Words and Lost Worlds: Modernity and the Language of Everyday Life in Late Nineteenth-Century Stockholm*, Cambridge: Cambridge University Press.
Pred, Allan (1990b) *Making Histories and Constructing Human Geographies: The Local Transformation of Practice, Power Relations and Consciousness*, Boulder, Colo.: Westview Press.
Pred, Allan (1991) "Spectacular articulations of modernity: the Stockholm Exhibition of 1897," *Geografiska Annaler*, 73B: 45–84.
Pred, Allan and Watts, Michael John (1992) *Reworking Modernity: Capitalisms and Symbolic Discontent*, New Brunswick, N.J.: Rutgers University Press.
Pred, Allan and Watts, Michael John (1994) "Heretical empiricism: the modern and the hypermodern," *Nordisk Samhällsgeografisk Tidsknift*, 19: 3–26.
Program och bestämmelser, Stockholmsutställningen 1928 (1928): (Stockholm: Bröderna Lagerström).
Råberg, Per G. (1972) *Funktionalistiskt genombrott – Radikal miljö och miljödebatt i Sverige 1925–31*, Stockholm: PA Norstedt & Söners förlag.
Råberg, Per G. (1980) "På väg mot en social estetik," in *1930/80 – Arkitektur Form Konst*, Stockholm: Kulturhuset, 6–16.
Rabinow, Paul (1986) "Representations are social facts: modernity and post-modernity in anthropology," in James Clifford and George E. Marcus (eds.) *Writing Culture: The Poetics and Politics of Ethnography*, Berkeley: University of California Press, 234–61.
Rabinow, Paul (1989) *French Modern: Norms and Forms of the Social Environment*, Cambridge, Mass.: The MIT Press.

Rawson, Judy (1976) "Italian futurism," in Malcolm Bradbury and Robert McFarlane (eds.) *Modernism 1890–1930*, Harmondsworth: Penguin.

Reslow, Wilhelm (1929) *En bokbinderiarbetares minnen*, Stockholm: Skrifter utgiven av Arbetarnas Kulturhistoriska Sällskap, II.

Reutersvärd, Oscar (1980) "Otto G Carlsund och hans postkubistiska expo," *1930/1980 – Arkitektur Form Konst*, Stockholm: Kulturhuset, 52–7.

Reutersvärd, Oscar (1988) *Otto G. Carlsund i fjärrperspektiv*, Åhus: Kalejdoskop förlag.

Reutersvärd, Oscar (1989a) "Det fanns en klyfta mellan revolutionernas utopister," *Dagens Nyheter*, March 3.

Reutersvärd, Oscar (1989b) "I revolutionens sfäriska salar," *Sydsvenska Dagbladet*, February 24.

Reutersvärd, Oscar (1990) *Franciska Clausen – De heroiska åren*, Stockholm: Boibrino Gallery.

Richards, J.M. (1940) *An Introduction to Modern Architecture*, Harmondsworth: Penguin.

Riksdagstryck (1895a) "Kongl. Maj:ts Nåd Proposition N:o 40."

Riksdagstryck (1895b) "Riksdagens protokol, Första Kammaren, N:o 29, Torsdagen den 2 maj."

Riksdagstryck (1895c) "Riksdagens protokol, Andra Kammaren, N:o 37, Torsdagen den 2 maj."

Riksdagstryck (1895d) "Statsutskottets Utlåtande, N:o 64, Ank. till Riksd. Kansli den 26 April."

Rosander, Göran (1989) "Dalkullor i reklamen," *Företagsminnen – Årsmeddelande från Föreningen Stockholms företagsminnen*, 22–33.

Rudberg, Eva (1981) *Uno Åhren*, Stockholm: Byggforskningsrådet.

Rudberg, Eva (1985) "Stockholmsutställningen 1930," *Daedalus – Tekniska Museets Årsbok*, 99–110.

Rudberg, Olle (1986) "Så gick det till på Per Froms Velocipid fabrik – Ur Fredrik Rydbergs dagbok från 1890-talet," *Sankt Eriks Årsbok*, 113–32.

Ruth, Arne (1984) "The second new nation: the mythology of modern Sweden," *Daedulus*, 113, Spring: 53–96.

Ryan, Michael (1989) *Politics and Culture: Working Hypotheses for a Post-Revolutionary Society*, Baltimore: The Johns Hopkins University Press.

Rydell, Robert W. (1984) *All the World's a Fair: Visions of Empire at American International Expositions, 1876–1916*, Chicago: University of Chicago Press.

Sahlin-Andersson, Kerstin (1989) *Oklarhetens strategi: Organisering av projektsamarbete*, Lund: Studentlitteratur.

Samec, Ann Charlotte (1989) "Globen ett 'medeltida samhälle,'" *Dagens Nyheter*, May 17.

Samuelsson, Kurt (1968) *From Great Power to Welfare State – 300 Years of Swedish Social Development*, London: George Allen and Unwin Ltd.

Sandberg, Nils-Eric (1992) "Som sömngångare mot monetär union," *Dagens Nyheter*, July 8.

Schlesinger, Philip (1991) *Media, State and Nation: Political Violence and Collective Identities*, London: Sage Publications.

Schorske, Carl (1981) *Fin-de-siècle Vienna: Politics and Culture*, New York: Vintage Books.
Scott, James C. (1985) *Weapons of the Weak: Everyday Forms of Peasant Resistance*, New Haven: Yale University Press.
Scott, James C. (1990) *Domination and the Arts of Resistance: The Hidden Transcript*, New Haven: Yale University Press.
Segel, Harold B. (1987) *Turn-of-the-Century Cabaret: Paris, Barcelona, Berlin, Munich, Vienna, Cracow, Moscow, St. Petersburg, Zurich*, New York: Columbia University Press.
Silverman, Debora L. (1977) "The 1889 Exhibition: the crisis of bourgeois individualism," *Oppositions: A Journal of Ideas and Criticism in Architecture*, 8 (Spring): 70–91.
Simmel, Georg (1896) "Berliner Gewerbe-Ausstellung," *Die Zeit* (Vienna), July 25.
Sjöblom, Anita (1991) "Tur vi har tennislirare," *Dagens Nyheter*, July 13.
Sjöström, Christer (1937) *En stockholmares minnen 1880–1900*, Stockholm: Studier utgivna av Arbetarnas Kulturhistoriska Sällskap, 6.
Söderberg, Rolf, and Rittsel, Pär (1983) *Den svenska fotografins historia 1840–1940*, Stockholm: Bonnier Fakta.
Sörenson, Ulf (1991). "Den fantastiska arkitekturen och förnuftet – Ferdinand Boberg – 90-talets arkitekt?,' *Artes*, 4: 139–54.
Sörenson, Ulf (1992) *Ferdinand Boberg – Arkitekten som konstnär*, Stockholm: Wiken.
Sorkin, Michael (1992a) "See you in Disneyland," in Michael Sorkin, (ed.) (1992b), 205–32.
Sorkin, Michael (ed.) (1992b) *Variations on a Theme Park: The New American City and the End of Public Space*, New York: Hill & Wang.
Sörlin, Sverker (1991) "Mellan mångfald och enhet: modernitet och identitet i Europa," in Sverker Sörlin and Jan-Erik Gidlund (eds.) *Ett nytt Europa*, Stockholm: SNS förlag, 9–64.
Ståhle, Carl Ivar (1981) *Stockholmsnamn och stockholmsspråk*, Stockholm: PA Norstedt & Söners förlag.
Stallybrass, Peter and White, Allon (1986) *The Politics and Poetics of Transgression*, Ithaca: Cornell University Press.
Stockholms kommunalförvaltning (1899) *Berättelse angående År 1897 jämte statistiska uppgifter för samma och föregående tid*, Stockholm.
Stockholms stads fastighetskontor (1934) *Bostadsförhållandena i Stockholm samt Stockholms stads bostadspolitik*, Stockholm: Beckmans boktryckeri.
Stockholms stads statistiska kontor (1930) *Statistisk årsbok över Stockholms stad 1930*, Stockholm: Beckmans boktryckeri.
Stockholms stads statistiska kontor (1931) *Statistisk årsbok över Stockholms stad 1931*, Stockholm: Beckmans boktryckeri.
Storper, Michael and Walker, Richard (1989) *The Capitalist Imperative: Territory, Technology, and Industrial Growth*, Oxford: Basil Blackwell.
Strömbom, Sixten (1965) *Nationalromantik och radikalism: konstnärsförbundets historia 1891–1920*, Stockholm: Bonniers.
Svedberg, Olle (1980) "Funktionalisms bostadsprogram – En bakgrundsskiss,"

Funktionalisms genombrott och kris – Svenskt bostadsbyggande 1930–1980, Stockholm: Arkitekturmuseet, 42–63.
Swedish Trade Council (1992) *Annual Report – Sweden 1991*, Stockholm: Swedish Trade Council.
Szabo, M. (1991) "Den nya konsumenten beträder scenen," *Fatuburen*, 130–54.
Tarschys, Daniel (1992) "Alliansfriheten består," *Dagens Nyheter*, August 11.
Tarschys, Rebecka (1991) "Kvinnohuset skapade mötet," *Dagens Nyheter* November 23.
Taussig, Michael (1993) *Mimesis and Alterity: A Particular History of the Senses*, New York: Routledge.
Taylor, Roger R. (1974) *The Word in Stone: The Role of Architecture in National Socialist Ideology*, Berkeley: University of California Press.
Tengdahl, Knut (1897) *Material till bedömande af hamnarbetarnes i Stockholm lefnadsförhållanden*, Stockholm.
Therborn, Göran (1981) *Klasstrukturen i Sverige 1930–1980 – Arbete, kapital, stat och patriarkat*, Lund: Zenit förlag.
Therborn, Göran (1988) "Hur det hela började – När och varför det moderna Sverige blev vad det blev," in Ulf Himmelstrand and Göran Svensson (eds.) *Sverige – Vardag och struktur*, Södertälje: Norstedts, 23–53.
Thomasson, Richard (1970) *Sweden: The Prototype of Modernity*, New York: Random House.
Thrift, Nigel (1992) "Muddling through: world orders and globalization," *The Professional Geographer*, 44: 3–7.
Thrift, Nigel and Corbridge, Stuart (eds.) (1994) *Money, Power and Space*, Oxford: Basil Blackwell.
Tomlinson, Alan (1990a) "Introduction: consumer culture and the aura of the commodity," in Tomlinson, Alan (ed.) *Consumption, Identity, and Style: Marketing, Meanings and the Packaging of Pleasure*, London and New York, Routledge, 1–38.
Tomlinson, Alan (ed.) (1990b) *Consumption, Identity and Style: Marketing, Meanings and the Packaging of Pleasure*, London and New York: Routledge.
Tunhammar, Göran (1992): "Vinklad EG-bevakning," *Dagens Nyheter*, July 7.
Turner, Victor W. (1969) *The Ritual Process: Structure and Anti-Structure*, London: Allen Lane.
Turner, Victor W. (1982) *From Ritual to Theatre: The Human Seriousness of Play*, New York: Performing Arts Journal Publications.
Unge, Ingemar (1992) *Paradiset under Globen*, Stockholm: Höjerings.
Velure, Magne (1988): "Djävla utlänning!," in Åke Daun and Billy Ehn (eds.) *Blandsverige – Om kulturskillnader och kulturmöter*, Stockholm: Carlssons, 182–205.
Verkställandeutskotts protokoll (1928–1930) minutes of the Executive Committee of the 1930 Stockholm Exhibition, held in SU–1930.
Virilio, Paul (1988) *Speed and Politics*, New York: Semiotext(e).
Von Horn, Paridon and Sundelöf, Fritz Gustaf (1975) *Staden sjunger: en Stockholms kavalkad*, Stockholm: Bokförlaget Prisma.
Wägner, Elin (1930) "Utopia," *Bonniers Månadstidning*, June, 20–3.

Wahlund, Helmer (1930) "Svea Rike," *Studiekamraten*, 13–14: 177–80.
Waldén, Katja (1980) "De voro alla ordets män," *1930/1980 Arkitektur Form Konst*, Stockholm: Kulturhuset, 35–49.
Wallén, Cyrus (1971) *En Södergrabb växer upp*, Stockholm: Nordiska museet.
Watts, Michael J. (1991) "Mapping meaning, denoting difference, imagining identity: dialectical images and postmodern geographies," *Geografiska Annaler*, 73B: 7–16.
Watts, Michael J. and Little, O.P. (1994) *Contract Farming in Africa*, Ithaca: Cornell University Press.
Werckmeister, O.K. (1991) *Citadel Culture*, Chicago: University of Chicago Press.
Westerberg, K. Martin (1930) "Stockholmsutställningen 1930," *Byggnadsvärlden*, 27A: June 30, 338–45.
William-Olsson, William (1937) *Huvuddragen av Stockholms geografiska utveckling 1850–1930*, Stockholm: Stadskollegiets utlåtanden och memorial, bihang 11.
Williams, Raymond (1977) *Marxism and Literature*, Oxford: Oxford University Press.
Williams, Rosalind H. (1982) *Dream Worlds: Mass Consumption in Late Nineteenth-Century France*, Berkeley: University of California Press.
Williams, Rosalind, H. (1990) *Notes on the Underground: An Essay on Technology, Society, and the Imagination*, Cambridge, Mass.: The MIT Press.
Willis, Paul (1990) *Common Culture: Symbolic Work at Play in the Everyday Culture of the Young*, Boulder: Westview Press.
Willis, Susan (1991) *A Primer for Daily Life*, London and New York: Routledge.
Wilson, Elizabeth (1992) "The invisible flaneur," *New Left Review*, 191: 90–110.
Wrede, Stuart (1980) *The Architecture of Gunnar Asplund*, Cambridge, Mass.: The MIT Press.
Zethelius, Otto (1899) "Utställningens folkkök," in Ludv. Looström (ed.) *Allmänna Konst – och Industriutställningen i Stockholm 1897. Officiel berättelse*, Stockholm, 192–98.
Zukin, Sharon (1991) *Landscapes of Power: From Detroit to Disney World*, Berkeley: University of California Press.

INDEX

Aalto, Alvar 100, 164
ABBA 214
Åbo (Turku) Exhibition of 1929 164
Ådalen 144
advertising and advertisements 56–8, 103, 117, 120–7, 131–3, 147, 158, 164–5, 200–1, 208, 213, 245, 247, 249
"advertising mast," the 117–28, 144, 196
Agnew, Jean Christophe 247
Albanians 219–20
Alingsås 127
Alu-Suisse 204
ANC (African National Congress) 216
Anderson, Perry 18, 240, 255
Andersson, Georg 220
Andersson, Henrik O. 109
Andersson, Roger 199
Andersson, Tomas 179
Andersson, Torbjörn 177, 188, 193
Andrée, S. A. 41
Anesäter, Stig 188
anti-Semitism 138–40, 154, 168, 221
Antwerp 67, 162
Armstrong, Louis 237, 255
Arp, Hans 147
Art Concret 146, 148, 173
ASEA 124

Asplund, Gunnar 98, 103, 110, 116, 124, 131, 138, 144, 147, 153, 155, 161, 163–5, 171, 181, 194
Åström, Sverker 217
Atlanta 90, 183
Austin, Alice Constance 172
Austria 217, 247
Austro-Hungary 72

Baker, Josephine 154
Bakhtin, Mikhail 249, 252
Baltic States 122
Baltimore 243
Baltimore Aircoil International 204
Barcelona 149, 162, 203
Baudrillard, Jean 88, 237–8
Bauhaus 130, 149, 165
Beatles, the 255
Becker, Boris 203
Bendix, Carl 84
Benedict, Burton 40
Benjamin, Walter 5, 11, 22–4, 29, 31, 51, 58, 89, 165, 247; *Passagen-Werk* (or Arcades Project) 11, 22–3, 46, 88
Bennett, Tony 51, 58, 75
Berg, Svante 245
Berg Arkitektkontor AB 193–4, 244
Berger, Peter L. 78

Berlin 24, 41, 45, 67, 72, 123, 149, 183; Dadaists in 29; Haus Germania in 166
Berman, Marshall 179
Bildt, Carl 200, 217, 220, 240
Björling, Jussi 214
Black Army 235–6
Blanche, Thore 84
Blomberg, David 133
Bloomfield, Terry 245
Blue Saints 235–6
Boberg, Ferdinand 90, 244
Boestad, B. G. 232
Bogart, Humphrey 237, 255
Bondi, Liz 198
Bonnier, Albert 144
Borg, Björn 214
Bosnia-Herzegovina 219–20, 251
Boston 243
Boulée, Etienne-Louis 182, 241
Boyer, M. Christine 66
Brandt, Willy 28
Broberg, Bengt 227
Brussels 73, 89, 182, 206, 244
Bryggman, Erik 164
Bubka, Sergei 203
Buck-Morss, Susan 11, 22, 46, 51, 89, 97
Buffalo 37
Bulgarian Turks 250
Bush, George 205

California 172, 244
Canada 123
capitalism(s) 13–14, 38, 110, 122, 188, 198, 203, 208, 232, 239, 255, 263–4; cultural creation under 18, 221; "flexible accumulation" and 204; inconstant geography of 185–7; industrial 32–4, 37
Carlsson, Ingvar 216
Carlsund, Otto G. 129, 145–8, 150, 156, 170–1
Casa del Fascio 196, 244

Castro, Fidel 249
Celik, Zeynep 47
Center Party 247
Central Bank of Sweden (*Riksbanken*) 212, 253
Certeau, Michel de 178
Charles, Ray 203, 216
Cher 203
Chicago 37, 45, 161; World's Columbian Exposition in 31, 39, 41, 51, 56, 73, 90; World's Fair of 1933 153
China 242
Christensen, Anna 200
citadel culture 177, 179, 187
Clark, T. J. 19
class tensions *see* social and class tensions and conflicts
Clausen, Francisca 147
Coca-Cola 183
Cold War 217–18
collective memory 66, 137, 155, 159, 194, 197, 202, 239, 261
commodity fetishism 13–14, 31, 51, 62, 81, 84, 154, 202
Communism 255
Como 196, 244
Conradson, Birgitta 12
Conservative Party *see* Moderate (Conservative) Party
constructivism 121, 130, 146–7, 164, 166, 195, 241
consumption 11–15, 19, 21, 34–5, 37–8, 46, 60, 66, 78, 81, 91, 100, 104, 106, 113, 116, 122, 130, 145, 150, 154–5, 157–8, 180, 182–3, 186, 197, 204, 206, 208, 210, 212, 223, 230–1, 234, 237, 248, 251, 259, 261–2; dreamworld of 47 ff., 82, 87, 93, 111–12, 258, 264; of commodified bodies 180, 202–3, 205, 208, 223; *see also* spectacle (the) and spectacular spaces of consumption

INDEX

Cooper, Alice 203
Copenhagen 251; Exhibition of 1888 89
Corso, Il 112–16, 124, 143, 151, 244
Count Basie Orchestra 202
Courier, Jim 203
creative destruction 14, 78, 83
Croatia 219–20
Cronon, William 51
Crystal Palace (Great Exhibition of the Works of Industry of All Nations) 37, 45, 75, 182
Cubism 144
cultural reworkings and cultural politics 220–1, 100, 150, 198, 223–31, 239, 253, 257–8, 262
Czechoslovakia 140

Dalecarlia 58, 63, 91
Danielsson, Randi 232
Davison, Graeme 40, 43, 45–6, 54
Debord, Guy 17, 89, 178, 198
Deep Purple 203
Denmark 43–4, 51, 69, 122, 213, 220, 246
Denning, M. 18
Depeche Mode 203
de Stijl 149, 165, 196
Detroit 161, 183
Disney Epcot Center 183
Disney on Ice 202
Djurgården (Stockholm) 41, 45, 77, 83, 101, 126, 227
Doesburg, Theo van 146–7
Domingo, Placido 203
Douglas, Louis, 138
Duncan, James 66
Dylan, Bob 202

Earth, Wind & Fire 202–3
Easter Island 237
Edberg, Stefan 203
Edinburgh 67
Edmonton 243
Eiffel Tower 59, 91–2, 120, 122, 164, 177, 192–3, 201, 237

Ekbom, Torsten 241
Eksteins, Modris 97
Ekström, Anders 56
electric light and illumination 64–6
Electrolux 124, 203
Eliasson, Axel 140, 169
Enders, Georg 133
England 46, 254
Englund, Peter 199
Engström, Albert 111
Environmental (Green) Party 245, 248
Eriksson, Eva 31
European Arena Association 203
European Community 16, 178, 181, 187, 199, 205–6, 211, 214, 216, 218, 233, 238, 245–9, 259, 261
European Union 16, 205–6, 257, 259
Ewen, Stuart 40, 85

Fälldin, Torbjörn 247
Farmers' Union Party (*Bondeförbundet*) 168
Fascism 112, 196–7
Featherstone, Mike 249
Federation of Swedish Industries (*Industriförbundet*) 206
Feldt, Kjell-Olof 246
Fijis, the (Fiji Islands) 237
Finland 46, 122, 204, 213, 220, 247, 251
Fischer, Michael M. J. 221
Fleetwood Mac 202
Ford Motor Company 183
Fordism 204
Foucault, Michel 11, 58, 66, 85
France 46, 98, 120, 184, 205, 220
French Revolution 67, 182
Freud, Sigmund 198
Friggebo, Birgit 220, 251
Frisby, David 47, 51
Frölich, Margareta Alexandra 156
Frykman, Jonas 116, 257
Fujisawa (Japan) 184
Fukuyama, Francis 240, 255

Fuller, Buckminster 182
functionalism 12, 104, 106, 117, 126, 134, 144, 147, 149, 153, 155, 157, 194, 258
functionalist architecture 98, 103, 106, 108, 110, 113, 116, 138, 144, 150, 151–4, 165
functionalist furniture and household goods 98, 103–4, 106, 108, 113, 150–1
futurism 112, 120, 163

Garbo, Greta 130, 200, 214
Geneva 251
Geneva Convention 219, 250
Germany 46, 72, 98, 108, 145, 205, 211, 218, 232
Giddens, Anthony 247
Glasgow 37
globalization and the global capitalist economy 175, 185, 191, 201, 204, 207, 221, 223, 233, 236, 238–9, 257
Globe, the 16, 19, 23, 175 ff., 259–62
Globen City 190–1, 193–4, 201, 209, 213, 225–6, 230, 232–3, 236, 239, 243, 252, 260
Goethe, Johann Wolfgang von 215
Göransson, Eva 177, 192
Göteborg 39, 127, 154, 160, 189, 224–5, 227, 251
Gradin, Anita 217
Graf Zeppelin 108
Gramsci, Antonio 18, 28
Great Barrier Reef 237
Great Exhibition of the Works of Industry of All Nations *see* Crystal Palace
Greco-Turkish War 44
Greeks 251
Green Party *see* Environmental (Green) Party
Gregory, Derek 11, 23, 257
Gren, Gunnar 214
Grevenius, Herbert 132

Gropius, Walter 112, 164
Grossberg, Lawrence 17
Grosz, Georg 29
Guérin, Charles 182
Guinness Book of Records 201
Guns 'n' Roses 203
Gustav Vasa *see* King Gustav Vasa

Hagia Sofia 43–4, 47
Hall, Stuart 18, 19, 179, 198, 221, 257
Hall, Thomas 188
Halland 200
Hamburg 89
Hamilton, Carl 240
Hammar MC 203
Haninge 231
Hansen, Per Albin 162
Haraway, Donna 85
Harning, Anderz 224
Harvey, David 188, 193, 222, 255
Hasselgren, A. 86
Hawaii 237
Heartfield, John 29
Hedlund, Olle 250
Hegel, Georg Wilhelm Friedrich 240
hegemonic discourses and representations 13, 17, 19–21, 24, 74, 82, 84–5, 134, 180, 183, 197–8, 201, 217, 230, 237, 261–4
hegemony and hegemonic interests 17–19, 25, 28, 84, 178, 188, 213–14
Heidenstam, Verner von 40
Heinrich Kamphöhler Gmbh & Co. 204
Hélion, Jean 144
Henny Penny 231
Hitler, Adolf 183
Höch, Hannah 29
Holiday on Ice 202
Holland 149
Holm, Hans Axel 226
Holmgren, Stefan 179
Horn, Paridon von 72
Hughes, Robert 13, 59
Hulth, Max 231

hypermodernity 15–17, 19, 21–2, 24, 110, 180, 202, 214, 237, 257–8, 262–4

IBM 203
identity 12, 17–18, 150, 157, 179, 223–4, 258; crises 17, 82, 100, 158, 181, 200, 202, 210, 211; European 207; individual and collective 15–19, 35–6, 81, 198, 235, 247, 262; national 40, 45, 70, 99, 121, 127, 140, 198–9, 213, 257
Iglesias, Julio 203
immigration *see* refugees and immigration
India 242
international exhibitions and World's Fairs 37, 40, 43, 45–7, 51, 54, 58, 70–2, 78, 87, 90, 92–3, 97, 100, 102, 162
Istanbul 43
Italy 160

Jackson, Janet 203
Jameson, Fredric 18
Japan 123, 204–5, 232
Jencks, Charles 244
Johanneshov (Stockholm) 189–90, 194, 196–7, 225, 243
Johansson, Gotthard 111
Johansson, Ingemar 214
John, Elton 202
Johnson, Eyvind 97, 110
Josefsson, Ingemar 189, 202, 227–8, 253
Josephson, Ernst 89
Jungstedt, Kurt 130

Källenius, Hans 240
Karling, Torsten 132
Karlström, Ulla 93
Key, Ellen 40, 48
King Gustav Vasa 67, 153, 237–8
King Oscar II 39, 57, 69, 71, 78, 82

Kirkeby, Anker 120, 130
kitsch 71–2, 83
Knutson-Tzara Greta 147
Köln Presse Exhibition 164
Kosovo 219–20
Kovacs, Josef 231
Kramfors 179
Kreisky, Bruno 28
Kreuger, Ivar 133, 142
Krinsky, Carol 244
Kuleshov, Lev 28
Kullager (Swedish Ball Bearing Co.) 133
Kylhammar, M. 98

L. M. Ericsson (Company) 124
Laclau, Ernesto 19
Lake Mälar 46
Larsson, Carl 89
Larsson, Yngve 31, 105
Leander, Zarah 108
Le Corbusier 11–12, 147–9, 160, 194
Ledoux, Claude-Nicolas 182
Left Party 222, 245, 248
Léger, Fernand 107, 145–7, 173
Lendl, Ivan 203
Lequeux, M.-J. 182
Lewerentz, Sigurd 165
Lewis, Carl 203, 216
Liberal Party 188, 192, 199, 231, 240
Liedholm, Nisse 214
Liljefors, Bruno 89
Liljeroth, Hans 189–90
Lilljekvist, Fredrik 90
Lindbergh, Charles 130
Lindgren, Astrid 214
Lissitzky, El 164
Lodin, Sven-Olof 217
Löfgren, Orvar 11, 12, 157, 200
Lo-Johansson, Ivar 109–11, 128, 149, 157
London 45, 182, 203, 243–4; *The Times* of 123
Louisiana 294
Luckmann, Thomas 78
Lund 69; Cathedral 201

Lundborg, Herman 132, 137, 168
Lundgren, Bo 178
Lundkvist, Artur 144
Lundqvist, Åke 231
Lynd, Helen Merrell 85
Lyttkens, Lorentz 97

Maastricht Treaty 216, 246, 249
Macedonians 250
MacLaine, Shirley 203
Makeba, Miriam 216
Maldives, the 237
Malm, Stig 199
Malmö 39, 69
Malmsten, Carl 117
Mandela, Nelson 216
Marcus, Georg 221
Marcus, Greil 178
Marx, Karl 198
Marxism 200–1
Medelsvensson, Mr. (Mr. Average Swede) 169, 218
MERO-Raumstruktur Gmbh 204
Meyer, Hannes 164
Miami 183, 237
Mies van der Rohe, Ludwig 112
Milan 203
Miller, Daniel 12, 251
Miller, Roger 244
Minnelli, Liza 203
Mitchell, Timothy 31, 40, 51, 67
Moderate (Conservative) Party 178, 189, 192, 199–200, 205, 219, 231, 233, 248
modernization and modernity 22, 24, 37, 42, 70, 78, 82, 99, 103, 109, 120, 150, 157–8, 215; capitalist 15, 22, 258, 262; European 21-2, 262–4; high 12, 15, 21–2, 24, 99, 109, 180, 194, 258, 261–3; industrial 13, 21, 24, 180, 258, 261–3; social 12–13, 28; Swedish 12–13, 15–16, 19, 117, 259, 263; *see also* hypermodernity
Moholy-Nagy, Làzslo 119, 146–7

Molière 215
Mondrian, Piet 97, 115, 146–7, 196
montage 22–7, 125, 262–4
Montreal World's Fair (1967) 182
Moscow 111, 121, 235
Moscow River 121
Munich 89, 203
Mussolini 112, 130, 196, 244

Näsström, Gustaf 111
National Institute of Racial Biology 137
NATO 217–18
Naylor, Gillian 98
Nehru, Jawaharlal 249
Neurath, Otto 134
neutrality, Swedish 100, 215–19
New Democracy Party 221, 248
New Kids on the Block 203
New Orleans 90, 243
Newton, Isaac 182
New York 97, 183, 187, 243; Metropolitan Museum of Art 101; World's Fair (1939) 171, 182; World's Fair (1964) 182
Nils-Georg 133
Nilsson, Birgit 214
Nilsson, Jan Olof 199
Nobel Industries 145
Nordahl, Gunnar 214
Nordström, Ludvig "Lubbe" 134–6, 167–8
Norrland 46
Norway 43–4, 51, 122, 145, 213, 218, 247
Nylén, Leif 111
Nyström, Bertil 134, 137

OECD 185–6
Ohlsson, Martin A. 31, 59
Old Town (Stockholm) 230
Old (*Gamla*) Stockholm at Stockholm Exhibition of 1897 67–71, 74–5, 79–80, 93–4, 260
Olsson, Axel Adolf 55, 73

Olsson, Gunnar 253
Opel 203
Örebro 69
Orlando 183
Osborn, Max 98
Oslo 203, 251
Östlundh, Håkan 193
Ozenfant, Amedée 147–8

Palme, Olof 191
Palmgren, Per 231
Pan American Airways 183
Pantheon, the 241
Paradise, the 112, 117, 128–33, 143–4, 153, 173
Paris 24, 31, 37, 41, 45, 58, 123, 145, 147, 149, 184, 187, 203, 236, 243; Exposition of 1867 93; Exposition of 1889 67, 73, 91–3, 182; Exposition of 1900 93, 182, 241; Exposition of 1925 101, 161
Patton, Phil 31, 56
Paulsson, Gregor 101–3, 106, 111, 117, 124, 130–1, 146–7, 150, 155, 161–2, 165–9, 171
Pavarotti, Luciano 203, 213
Persian Gulf War 216
Persson, Kristina 199
Peterson-Berger, Vilhelm 120
Philadelphia 37
Phuket 237
Pioneer (Inc.) 203
Poitiers 184
Poland 122
Pomian, Krzystof 178
Pope John Paul II 213
Portal House 196–7
post-Cubism 145, 147
postmodernism and postmodernity 15, 23, 46–7, 69, 190, 192–8, 238, 244, 253
Potsdam 170
Prague 183, 241
private space 208–9, 243

privatization of welfare and cultural institutions 16, 210–11
public space 208–9, 247
Puerto Vallarta 237
purism 147
Pyramide du Louvre 192

Queen Sofia 43

Råberg, Per G. 98, 110
Rainbow, Paul 28
racism 137–40, 154, 159, 168, 221, 251, 257
Reagan, Ronald 205
Reclus, Elisée 241
refugees and immigration 219–21, 250, 257
Reutersvärd, Oscar 170, 241
Rexed, Knut 199
Richard, Cliff 202
Richards, J. M. 98
Rolander, Dag 240
Rolling Stones, the 255
Rome 241
Ross, Diana 202
Rousseau, Henri 237–8
Roxette 203
Russia 43–4, 46, 51, 121, 127, 218
Russian Revolution 241
Ryan, Michael 18
Rydberg, Lennart 231–2
Rydberg, Viktor 137
Rydell, Robert W. 78

Sachs, Josef 140
St. Louis 37
St. Petersburg 145
Scandinavium 189, 224, 251
Schein, Harry 251
Schwitters, Kurt 13
Scott, James C. 85
Seattle 243
Separator (Inc.) 133
Serbs 219

Serge, Monsieur 120
Seville World's Fair 214
Shakespeare, William 215
Shand, P. Morton 98
SIAB 189
Silverman, Debora 45, 67
Simmel, Georg 47, 51
Simon, Paul 202
Simple Minds 203
Sinatra, Frank 203
Situationist International 88
Sjöberg, Patrik 203
Sjöblom, Anita 200
Skåne 22
Snoilsky, Count Carl 43–4
social and class tensions and conflicts 13, 40, 58, 76, 78, 91, 131, 142–4
social democracy 21
Social Democrats and Social Democratic Party 13, 15–17, 28, 86, 99, 106, 150, 157–8, 162, 168, 189, 192, 199, 209, 211–12, 216–17, 219–20, 225, 228, 230, 233, 243–5, 247–8, 250–1, 259, 262
social engineering 17, 99, 106, 138, 150, 154, 157–9, 168, 212, 258, 262
Södermalm (Stockholm) 189, 222, 251
SONY (Inc.) 203–4
South Africa 216
Soviet Union 203, 247
spectacle (the) and spectacular spaces of consumption 17, 19, 21–2, 24, 37, 45, 51, 58, 66, 71, 75, 84, 122, 178, 180, 192, 194, 223, 225, 237, 239, 243, 259, 263
Speer, Albert 183
Springsteen, Bruce 203
Stalingrad 218
Stallybrass, Peter 249
State Church (Swedish) 137, 168
Stewart, Rod 203
Sting 203
Stockholm 15, 22, 24, 31–2, 34–6, 45, 53, 73 ff., 102 ff., 179 ff.; City Hall 201; Exhibition of 1897 13, 19, 23, 31 ff., 11, 154, 164, 239–40, 244, 258–62; Exhibition of 1930 12–13, 19, 23, 97 ff., 181, 193–6, 233, 239, 258–62; Globe Arena *see* Globe, the; Globe Hotel 196, 208, 226, 234, 244, 247; housing conditions in 104–5, 189, 225; Propaganda Inc. 240; *see also Djurgården, Johanneshov*; Old Town; Old (*Gamla*) Stockholm at Stockholm Exhibition of 1897; *Södermalm*
Strindberg, August 47
Sukarno, Akmed 249
Sundelöf, Fritz Gustaf 72
Sundgren, Per 22
surrealism 144, 146
Svea Rike 117, 132–44, 153, 167, 169–70, 172, 260–2
Swedish Air Traffic Department 136
Swedish-America Line 173
Swedish Confederation of Labor Unions (LO) 17, 102, 199
Swedish Employers' Confederation 178
Swedish Flag Day 226
Swedish Handicrafts Association 101–2, 160
Swedish National Immigration Authority 220
Swedish National Railways Board 136
Swedish Post Office Administration 136
Swedish State Power Board 136
Swedish State Railways 218
Swedish Telegraphic Service 136
Swedish Trade Council 231, 253
Swedish Union of Commercial Employees 224, 251
Switzerland 160, 247
Sydney 243
symbolic discontent 25, 32–3, 81, 157, 198, 225, 253

taboo-breaking and silence about 85–7
Tamm, Baron Gustaf 39, 44, 79, 82, 84, 87

INDEX

Tarde, Gabriel 92
Tatlin, Vladimir 121
Täuber-Arp, Sophie 147
Taussig, Michael 85
Taylor, B. 193
Terragni, Giuseppe 196, 244
Thatcher, Margaret 205
Therborn, Göran 97, 159
Third International 121
Thule 44
Tito, Marshal 249
Tomlinson, Alan 12
Toronto 184, 243
Tranås 148
Tunhammar, Göran 178
Turks 251
Turner, Tina 202
Turner, Victor W. 247

unemployment 142, 169, 233, 253, 257, 260
Unge, Ingemar 222, 240
United Kingdom 202, 205, 220
United Nations 216–17, 251; High Commission for Refugees 220
United States 41, 46, 97, 123, 182, 184, 202, 205, 213, 216, 218, 232, 243, 250
Uppsala 121, 137; Cathedral 201

Vadstena 69
VAM (*Vit Arisk Motstånd*, or White Aryan Resistance) 221, 236
Vandermaelen, Phillippe 182
Verdi's Requiem 213
Vesnin brothers (A. and V.) 121, 164

Vienna 37, 72, 203
Villard-Cotard Globe 182
Visby 69
vision, technologies of 38, 58–61, 81, 134–5, 183
Volvo (Inc.) 203–4, 251

Waern-Bugge, Ingeborg 152
Wägner, Elin 144, 157
Wallén, Cyrus 54
Wallenberg, Knut 102
Wallenberg, Marcus 102
Walloons 215
Wall Street crash of 1929 142
Warsaw Pact 217
Watts, Michael J. 18, 19, 29, 175
Weimar Republic 164
Werckmeister, O.K. 175, 177, 187–8, 194
Westerberg, Bengt 220
White, Allon 249
Wiklund, Rolf 232
Williams, Raymond 18
Williams, Rosalind 46, 61, 92
Willis, Paul 12, 18, 221, 251
Willis, Susan 23
World's Fairs *see* international exhibitions and World's Fairs
World War I 97, 135
World War II 199, 218
Wyld's Great Globe 182

Yugoslavia 219–20, 250–1

Zorn, Anders 89
Zukin, Sharon 177